AUTOMOBILE FAULT-TRACING

PITMAN'S AUTOMOBILE MAINTENANCE SERIES

AUTOMOBILE FAULT-TRACING

A PRACTICAL HANDBOOK FOR SERVICE MECHANICS AND OWNER-DRIVERS

BY

STATON ABBEY, M.I.M.I.

Author of
"Automobile Workshop Practice"
"Diesel Fault-tracing, Maintenance and Repair"
"Automobile Transmission Servicing and Overhaul"
"Automobile Steering, Braking and Suspension Overhaul," etc.

THIRD EDITION

PITMAN PUBLISHING

First published 1951
Reprinted 1955
Reprinted 1957
Second edition 1961
Reprinted 1962
Third edition 1966
Reprinted with minor amendments 1968
Reprinted 1970

SIR ISAAC PITMAN AND SONS LTD.
Pitman House, Parker Street, Kingsway, London, WC2B 5PB
P.O. Box 6038, Portal Street, Nairobi, Kenya

SIR ISAAC PITMAN (AUST.) PTY. LTD.
Pitman House, Bouverie Street, Carlton, Victoria 3053, Australia

PITMAN PUBLISHING COMPANY S.A. LTD.
P.O. Box 9898, Johannesburg, S. Africa

PITMAN PUBLISHING CORPORATION
6 East 43rd Street, New York, N.Y. 10017, U.S.A.

SIR ISAAC PITMAN (CANADA) LTD.
Pitman House, 381–383 Church Street, Toronto, 3, Canada

THE COPP CLARK PUBLISHING COMPANY
517 Wellington Street, Toronto, 2B, Canada

ISBN: 0 273 40233 1

MADE IN GREAT BRITAIN AT THE PITMAN PRESS, BATH
G0—(T.4004)

PREFACE

FAULT-TRACING can be either an absorbing or an exasperating process, depending on how it is tackled. In this book the practical approach is taken, and the importance of a logical and systematic method of diagnosis is stressed. AUTOMOBILE FAULT-TRACING is intended to help the average owner as well as the experienced mechanic. Since the majority of readers will not have access to the more elaborate fault-tracing equipment, wherever possible tests have been described which can be made without using specialized instruments.

Each chapter contains a number of key charts, into which is concentrated all the essential information required to diagnose a particular fault ; the various tests and their implications are discussed in the text. Where possible the appropriate remedy is indicated, although often the cure is self-evident once the fault has been isolated.

In this respect it has been difficult to decide, in some instances, just where to draw a firm line of demarcation between fault-tracing and repair or overhaul. In a book in which the emphasis is on fault-tracing, obviously all available space should be devoted to this subject : yet this in itself is of little value unless some indication is given of the method of remedying a trouble once it has been traced. A balance has been struck, therefore, by detailing the simpler methods of rectifying faults, including adjustments and " first-aid " measures, and indicating when major tasks, such as engine, transmission, or chassis overhauls, are required.

A book of this type could not claim to be comprehensive without drawing on the assistance and experience of the service departments of the major car manufacturers and of the makers of many specialized components. Thanks are therefore due to a number of firms who have co-operated in this respect. Among these are: The British Motor Corporation, General Motors, Automotive Products Company, Delco-Remy, the Rootes Group, Vauxhall Motors, the Ford Motor Company, and the Dunlop Rubber Company.

In this respect it is only fair to mention that where specific faults are dealt with, they are by no means peculiar to, or associated with, any one make of car, but are representative of troubles which may be experienced over a wide range of vehicles, usually as the result of neglect or misuse in one form or another.

In each new edition since this book was first published it has been necessary to carry out extensive revisions, to keep pace with the progressive development in automobile design and technology. A good deal of information dealing with early models has been retained, however, as it is appreciated that the older the car, the more likely is a book of this type to be of real value to a practical owner or a mechanic.

Finally, the comprehensiveness of succeeding editions will depend, to a certain extent, on the co-operation of practically-minded readers, who may have devised useful methods of tracing and rectifying elusive faults. I shall always be pleased therefore to receive criticisms and suggestions from which other readers may benefit in later editions.

ST. OSYTH
ESSEX

STATON ABBEY

CONTENTS

FAULT-TRACING CHARTS

CHAPTER I

THE PRINCIPLES OF FAULT-TRACING

FAULT-TRACING may be defined as both an art and a science : an art, because skill, experience, and what can sometimes be described only as intuition, will enable some owners and mechanics to diagnose a trouble with the minimum expenditure of time and effort ; and a science, in that successful diagnosis of more elusive faults calls for logical deduction and, sometimes, the assistance of testing equipment and instruments.

Logical analysis of the problem, however, is the major factor in successful fault-tracing. Occasionally, a lucky guess at the cause of the trouble may be correct : a glance at Chart No. 1, however, will indicate that there are more than thirty-five main causes of difficult starting at normal temperatures, while Chart No. 3 details nearly fifty faults which may cause misfiring and loss of power. The chances of striking the correct item at the first attempt are, therefore, slender. When the main items in the two charts quoted are analysed in much greater detail in other charts, the need for systematic diagnosis becomes evident.

It will already be obvious, however, that if logical fault-tracing principles are to be applied to all the components which make up the modern car, taking into consideration all interdependent and correlated faults, the subject is apt to become extremely complex. To overcome this difficulty the special lay-out of the charts in this book has been adopted. Into these it has been possible to compress an amount of information which would otherwise occupy several volumes. The text, moreover, can be devoted to a fuller explanation of any points which need clarification.

This chapter will serve as an introduction to the subject and as a general survey of the principles employed.

Fault-tracing Equipment. Engine testing and tuning equipment has been developed to a considerable extent during recent years, and the elaborate test panels installed in many service stations considerably simplify the analysis of engine and electrical system faults. These test sets comprise some or all of the following basic instruments—

A *vacuum gauge*, to measure the depression in the engine induction system, in itself one of the most reliable guides to engine efficiency.

A *compression gauge*, with which the compression pressure in the cylinders is measured : this enables cylinder, piston, ring and gasket troubles to be quickly diagnosed.

A *pressure gauge*, used for fuel pump testing.

A *voltmeter* and *ammeter*, for electrical checks.

A *standard spark gap*, sometimes termed a " sparkmeter," used to check the ignition high-tension current voltage.

A *condenser tester*, to check condenser capacity.

An *ohmmeter*, useful in measuring resistances in electrical circuits.

A *motor-driven contact breaker*, which enables the performance of ignition coils to be checked, and, in addition, a *motor-drive* by which the distributor taken from the car can be driven while its performance is checked at various speeds.

An *exhaust gas analyser*, invaluable for carburettor tuning.

In many modern engine analysers, electronic circuits replace some of the above instruments and the results of the tests (which can be carried out with the engine running) are displayed on the screen of a cathode-ray tube.

While the foregoing may seem a formidable list, only vacuum and compression gauges, a voltmeter, and an ammeter (if the latter is not fitted as standard to the car) can be regarded as necessary for most of the tests described in this book, and even these instruments can be dispensed with in many instances. When an actual voltage reading is not required, for example, as when checking the continuity of wires in electrical circuits, a small lamp of battery voltage, in a holder to which are attached flexible leads fitted with metal test prods, will serve very well. If one test prod takes the form of a stout darning needle, the test bulb can be used to check the continuity of an insulated wire at short intervals along its length by pressing the point of the needle through the insulation ; the exact point at which the wire is fractured can be detected in this manner.

Road Tests. Since the uses of the various engine and electrical testing devices are fully covered in the appropriate chapters, attention can be turned to the principles of road testing, which may supplement engine testing and tuning, or

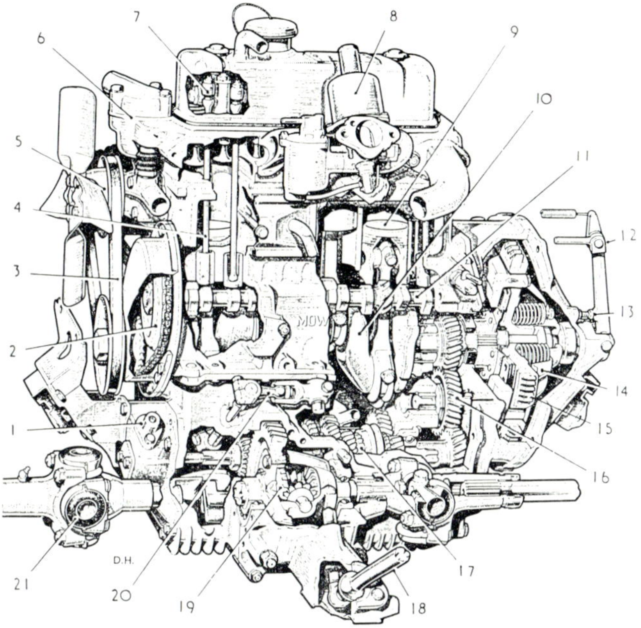

FIG. 1. TRANSVERSE ENGINE AND GEARBOX ASSEMBLY OF B.M.C. MINI MODELS

The arrangement of the B.M.C. 1100 and 1800 engines is similar. Modern power units are complex affairs—hence the need for systematic diagnosis of faults

1. Speedometer drive
2. Timing chain and sprockets
3. Fan and generator driving belt
4. Valve push-rod
5. Fan and water pump
6. Thermostat housing
7. Valve rocker
8. Carburettor
9. Piston
10. Crankshaft
11. Camshaft
12. Clutch-operating arm
13. Clutch adjustment
14. Clutch
15. Starter ring gear on flywheel
16. Idle gear between crankshaft and transmission gears
17. Transmission gears
18. Gear lever
19. Final-drive and differential gears.
20. Gear selector
21. Inner universal joint on driving shaft

which may be carried out to detect transmission, chassis, and body troubles. Again a logical approach is essential, determining, for instance, whether a fault is apparent at all speeds or only at high or low speed ; when the engine is driving the car, or on the over-run ; in top gear, or in only one gear, or when the car is coasting with the gear lever in neutral and the engine stopped. Examples of this type of analysis will be found in Chapters VII to XII.

When the road test is carried out as a check on engine performance, it should be conducted over a closed road circuit

FIG. 2. A MAGNETIC CRACK DETECTOR

This type of detector, generally found only in large workshops, detects cracks and internal defects which, even under strong magnification, are otherwise invisible

which includes a variety of gradients and road surfaces. If timed acceleration and speed tests are carried out, they should be made in both directions over the same stretch of road to cancel out favourable and adverse factors of wind, road surface, and gradient.

Petrol Consumption Tests. When making fuel consumption tests this is particularly important, while the need for accurate measurement of the relatively small amounts of fuel consumed during the tests calls for the use of a separate test tank which holds, say, a pint or a quart of petrol. Although the graduated measuring tanks or flow meters used by carburettor service stations represent the most accurate method of measuring the petrol, adequate results can be obtained by making up a measuring tank from a quart oil tin, into the base of which is soldered a union to take a flexible petrol pipe by which the tin is connected to the inlet union of the petrol pump, the suction line to the car tank being temporarily disconnected and blanked off.

The tank should be filled to a level mark before the test, and the exact amount of petrol required to bring the petrol level back to the mark at the conclusion of the test should be determined with a graduated measuring glass of the type obtainable from most chemists. Although this may seem an unnecessarily complicated process, it gives more accurate results than does relying on an arbitrary amount of petrol contained by a tank with a nominal capacity of, say, a pint or a quart—a capacity which may or may not be accurate.

When conducting petrol consumption tests it is usual to check the distance run by means of the speedometer mileage recorder. It is, therefore, essential to check this for accuracy beforehand. Similarly, acceleration and speed tests which rely on the accuracy of the speedometer speed-indicating mechanism are liable to be considerably upset by faults in the speedometer driving cable or speedometer itself. Some aspects of speedometer testing are covered in Chapter XIII.

Measuring Engine "Pull" and Tractive Resistance. In order to measure the "pull" of the engine and the tractive resistance set up by the transmission, wheel bearings, tyres and possibly slightly binding brakes, the well-known Tapley or Bowmon meters are invaluable, providing a direct means of comparison of the effects of different carburettor settings, ignition timing, and other alterations which affect engine efficiency under various running conditions.

The advantage of carrying out tests with the aid of a meter of this type is that variable factors such as errors in timing, differences in wind force, road surface, gradient and so on, are ruled out. The meter makes use of a pendulum, geared to a rotating scale. As the car accelerates, the pendulum weight is, as it were, left behind; the scale is rotated, and the rate of

INSPECTION
FAULT TRACING AND

/EHICLE MAKE AND YEAR .... CAPACITY OR HORSE POWER .... REG. NUMBER

ENGINE AND ACCESSORIES

Compression ....

Tappets : clearance ; noise ....

Valve Gear : noise ; cover gaskets ....

Timing Gears or Chain : noise ; backlash...........

Cylinder Head : nuts ; gaskets ....

Manifolds : securing nuts ; gaskets ; air or gas leaks...

Driving Belts : fan ; dynamo ; accessories ....

Water Pump : leakage ; effectiveness ....

Radiator : water level ; leakage ; filler cap ; condition of core ; shutters ; relief valve ; overflow pipe ....

Oil Leaks : crankcase ; pipes and unions ; timing covers ; main bearings ; drain plugs ....

Oil Filters : sump ; external ....

Engine Mountings ....

CARBURETTOR AND FUEL PUMP

Jets ....

Float Chamber Level ....

Choke: thermostat setting...............................

Easy Starting Device : interconnecting linkage ; thermostat ; dashboard control ....

Fuel Pump : effectiveness ; delivery pressure......

Filters : carburettor ; petrol pump ; petrol tank ; air cleaner ....

IGNITION SYSTEM

Contact Breaker Points : gap ; condition ....

Rotor : condition ....

Distributor Cap : cleanliness ; cracks ; tracking...

Timing Control : manual ; centrifugal ; suction operated ....

Coil and Wiring : efficiency ; condition of high and low tension wiring ; support....................

Magneto : efficiency ; impulse starter; earthing wire ....

Sparking Plugs : gap at points ; condition of points and insulators ; washers ....

ELECTRICAL SYSTEM

Dynamo or Alternator: charging rate..................

Dynamo brushes and commutator : burning or pitting ; chatter and sparking........................

Dynamo or Alternator Drive: condition of belt; slack in chain or gear...................................

Voltage Regulator ....

Cut Out and Fuses ....

Ammeter : accuracy ....

Battery : acid level ; specific gravity, voltage of each cell ; connexions ; corrosion ....

LIGHTS

Headlamps : focus and alinement ; condition of reflectors ....

Side lamps ....

Tail Lamps..

Stop Lamp ....

Other Lamps : ignition warning ; interior ; trafficators...

Switches and Fuses ....

ELECTRICAL ACCESSORIES

Windscreen Wipers : efficiency ; noise ; condition of blades ....

Trafficators ....

Starter : efficiency ; condition of pinion and flywheel teeth...

Horns ....

Interior Heater ....

Defroster ....

Radio ....

STEERING

Lost Motion...

Stiffness ....

Front Axle Alinement : toe in ; castor and camber angles ....

King Pins and Bushes ....

Front Hub Bearings ....

Steering Linkage and Joints............................

RECORD

CONDITION REPORT SHEET

.... SPEEDOMETER READING................. DATE OF INSPECTION...................

SUSPENSION

Springs: condition of leaves; broken leaves; clips and shackles ....

Shock Absorbers: adjustment; fluid level; security of attachments..

Torque Reaction Rods or Cables: adjustment; security of attachment ....

BRAKES

Pedal Travel ....

Condition of Rods or Cables ....

Hydraulic System: fluid level; leakage from pipe lines; choked pipes; air in system.................

Handbrake: adjustment; condition of ratchet...

Servo Mechanism: vacuum or air cylinder; hydraulic booster pump; valve mechanism; accumulator; pipe lines ....

TRANSMISSION

Clutch: adjustment ....

Backlash: in universal joints; in back axle; in gearbox ....

Rear Axle: leakage ....

WHEELS

Rims: buckling; rust ....

Spokes: buckling; looseness ....

Wheel Nuts: tightness; condition of studs; splines on knock-off wheels ....

TYRES

Pressures ....

Tread Wear ....

Condition of Sidewalls ....

Valves and Caps...

BODYWORK

General Appearance ....

Windows: winding controls; glass; channels......

Doors: alinement; buffers; locks; check straps..

Sunshine Roof: operation; water leaks ....

Seats: condition of springs; adjustments ....

Rear Blind ....

Bumpers: condition; attachments ....

Tool Kit ....

Fire Extinguisher ....

ROAD TEST

ENGINE

Ease of starting..................

Oil pressure when hot ....

Idling ....

Acceleration ....

Maximum Speed ....

Fuel Consumption...............

Oil Consumption ....

Engine Temperature............

Exhaust: noise; restriction; blue smoke.....................

General Noise.....................

Fumes.............................

CLUTCH

Slip ....

Fierceness ....

Spinning ....

Unusual noises ....

GEARBOX

Noisy gears.......... ....

Synchromesh operation ......

Preselector operation ....

Vibration on Gear Lever ......

Gears Jumping out of mesh...

PROPELLER SHAFT

Vibration ....

Noise ....

BACK AXLE

Noise on Drive ....

Noise on Overrun ....

Noise on Corners ....

INSTRUMENTS

Functioning ....

Accuracy ....

SUSPENSION

Comfort ....

Roll on corners ....

BRAKES

Power ....

Balance ....

Smoothness ....

Judder or Squeal ....

Binding ....

CHASSIS

Whip ....

Rattles ....

STEERING

Positiveness ....

Castor action ....

Wander ....

Road Shocks at Steering wheel ....

acceleration is recorded. The same thing happens when the car is climbing a hill. Conversely, deceleration, such as is caused by tractive friction when the car is running freely on the level with the gear in neutral, swings the pendulum weight forward and rotates the scale in the opposite direction.

The reading is shown in lb per ton in order to provide an indication of performance which is independent of the actual weight of the car. A reading of 175 lb per ton, for instance, represents exactly the same rate of acceleration or hill climbing ability on a car weighing one ton as on one weighing five tons ; such a pull will cause any vehicle to accelerate at $1\frac{3}{4}$ m.p.h. during every second that the pull is maintained on a level road, or will enable it to climb a gradient of 1 in $12\frac{3}{4}$ without losing speed. The maximum pull is generally recorded between 20 and 30 m.p.h. in top gear, the actual figure depending on the type of car and its mechanical condition : it may be between 100 and 300 lb per ton.

Tractive resistance should be measured with the gear lever in neutral at a speed of 10 m.p.h. on a good, level road surface, making a test in both directions if a strong wind is blowing. When a car is fully run-in, the resistance should not exceed 35 lb per ton, although a higher figure is usually recorded in winter if a test is made soon after a start from cold, owing to the greater viscosity of the oil in the rear axle and gearbox.

Preventing Trouble and Restoring Performance. There is a great deal to be said for carrying out a systematic inspection of the car at regular intervals—say, every 5,000 miles—in order to detect any incipient trouble and to correct any faults which may exist. The thoroughness of the inspection will naturally depend on the time which can be spared and the testing equipment available. A suggested routine is given in the fault-tracing and condition report sheet on pages 6 and 7. This covers all essential items, and, if it is conscientiously carried out and any necessary adjustments are made, it will restore " new car " performance on a vehicle that is in reasonably good mechanical condition. Any major faults which are present, such as loss of engine compression, and transmission or similar troubles, will, of course, require attention as soon as possible.

An inspection record of this type has, in itself, a number of advantages. Not only does it ensure that no item is overlooked, but it also enables the general condition of the car to be assessed at a glance. If completed systematically at each inspection period and filed for reference purposes, it will serve as a guide to the reconditioning required and as a check on the performance of each item. When made out for the first time, as with a newly-purchased used car, the chart is almost certain to reveal defects that might otherwise have been unsuspected and that might have developed into major troubles.

CHAPTER II

ENGINE: OPERATIONAL FAULTS AND TUNING

PROBABLY the best example of a logical series of tests is the process of diagnosing the cause of sudden stoppage of the engine during normal running. Often, of course, the presence or absence of premonitory symptoms will give some guide to the cause of the stoppage. It is not logical, for instance, to attribute a sudden and complete cessation of activity to the carburettor: if carburation trouble or shortage of fuel is to blame, misfiring or spitting back will almost certainly precede the final stoppage. An ignition fault, on the other hand, such as a low tension wire becoming detached from the coil, or the high tension lead between the coil or distributor coming adrift, will often cause the engine to stop suddenly. A broken ignition contact breaker spring or a detached battery lead will have the same effect.

When in doubt, however, or when called upon to examine a car, the previous history of which is unknown, the quickest course is to follow the sequence of tests shown in Chart No. 1.

DIFFICULT STARTING AT NORMAL TEMPERATURES

The most baffling problems are usually encountered when an engine stops or refuses to start at normal atmospheric temperatures, although there is no immediate obvious cause of difficult starting, and the starter motor rotates the crankshaft satisfactorily, or the engine can be rotated by means of the handle, or by towing, at a speed which normally is sufficient to ensure an easy start. By following the sequence of tests given in Chart No. 1, however, it should be possible to track down the trouble without delay.

It may be mentioned that since it will probably be necessary to carry out this work without the aid of specialized test equipment, as for example in a roadside breakdown, only simple and straightforward tests have been included. The use of valuable aids such as compression and vacuum gauges, fuel pump analysers, ignition test equipment, and so on, is more appropriate to the tracing of loss of power and misfiring, as described later in this and subsequent chapters.

Check the Petrol Supply. Although the petrol gauge may indicate that there is petrol in the tank, this instrument is by no means infallible. The actual presence of petrol should be

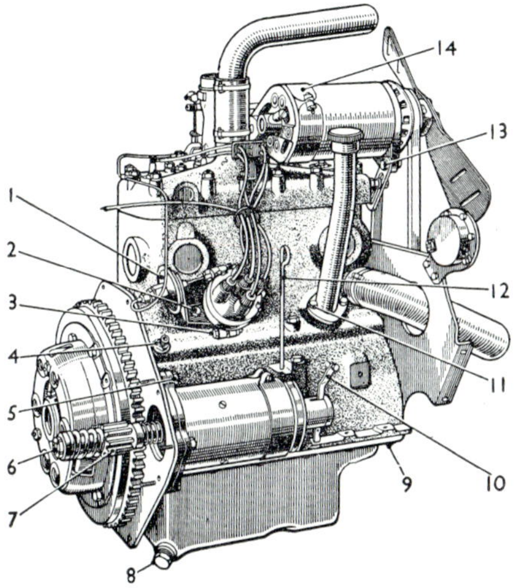

FIG. 3. TYPICAL SIDE-VALVE POWER UNIT, SHOWING FAULTS WHICH MAY CAUSE DIFFICULT STARTING, MISFIRING, OVERHEATING AND LUBRICATION TROUBLES

1. Vacuum ignition timing control inoperative or incorrectly adjusted
2. Loose distributor locating bolt
3. Slack distributor clamping bolt ; incorrect igniton timing
4. Oil leakage at oil gauge pipe union due to looseness or defective gasket (Chap. III)
5. Starter motor securing bolts loose
6. Starter motor shaft bent ; nut loose ; spring broken
7. Starter pinion binding on sleeve
8. Leakage from oil sump drain plug
9. Oil sump securing studs loose, or gasket defective (Chap. III)
10. Starter switch defective, contacts burnt, control wire incorrectly adjusted
11. Oil filler flange screws loose
12. Oil level in sump too low : check dipstick reading
13. Dynamo and fan belt tension incorrectly adjusted
14. Dynamo commutator and brushes dirty, scored, or worn

checked if possible by means of an improvised dipstick, such as a jack handle. The filler pipe may be curved, however, or may contain baffles which render this impossible.

The quickest method, therefore, is to flood the carburettor (when a plunger is fitted to the float chamber cover for this purpose), or to remove the float chamber cover or bowl to make sure that the float chamber is full of petrol. When an electrical pump is fitted the ignition must, of course, be

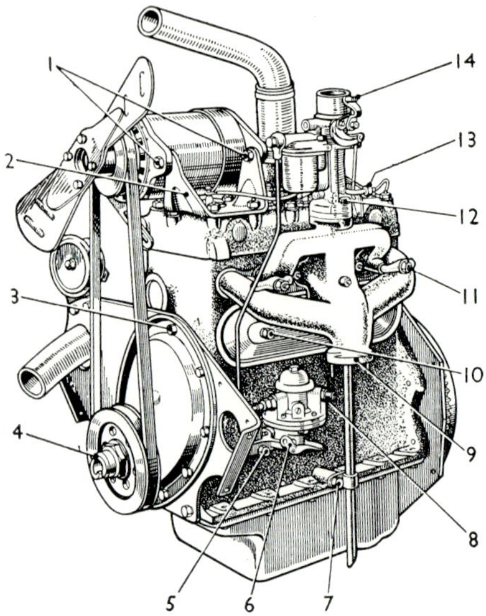

FIG. 4. THE NEARSIDE OF THE ENGINE SHOWN IN FIG. 3, ILLUSTRATING MORE COMMON FAULTS

1. Dynamo mounting bolts loose
2. Dynamo bracket loose on cylinder head
3. Loose timing chain cover screws or defective gasket, allowing oil leakage (see Chap. III)
4. Worn starting handle dog
5. Petrol pump mounting studs loose
6. Hand priming lever inoperative
7. Loose oil breather pipe clip ; clogged breather pipe
8. Loose petrol unions. Petrol pump defective. Filters clogged
9. Gas leakage at exhaust pipe flange
10. Valve cover securing screws loose or gasket defective, allowing oil leakage
11. Manifold bolts loose, allowing gas or air leakage ; gaskets defective ; manifolds not correctly lined-up with ports
12. Air leakage at carburettor flange : loose nuts ; defective gasket ; distorted flange
13. Vacuum ignition control pipe union loose, allowing air leakage to carburettor and upsetting action of ignition control
14. Carburettor controls incorrectly adjusted

switched on while flooding the carburettor ; if the pump buzzes, lack of petrol or an air leak in the fuel pipe between the pump and the petrol tank can be diagnosed immediately. A mechanical pump should be primed by means of the hand priming lever, if fitted, or by rotating the engine a number of times with the starter or handle. Failure of the pump to supply petrol calls for the action shown on the chart, followed, if necessary, by reference to Chapter V.

Carburation or Ignition? If the carburettor can be flooded or the float chamber is full, we are faced with our first major alternative: does the trouble lie in the carburettor or ignition system? The next step in the process of elimination is, therefore, to remove one sparking plug and rest it on the cylinder head with its body in contact with bare metal but its terminal well clear. See that the ignition is switched on, and rotate the engine.

If a spark occurs and the points and internal insulator of the plug appear clean, the insulator being a light biscuit colour, it is safe to assume that the ignition system is satisfactory. Continue with the tests given in Chart No. 1.

These are self-explanatory, except for the fact that on overhead valve engines it may not be easy to introduce petrol into the cylinders through horizontal plug holes unless a syringe is available; in these cases it is best to investigate the various carburation and induction faults described in Chapter V or the mechanical faults covered in Chapter III.

It is as well, however, to make sure that the trouble is not due to an over-rich mixture, caused by excessive use of the choke or starting control when the engine is warm, or by repeatedly depressing the accelerator pedal when an acceleration pump is fitted to the carburettor. Open the throttle fully and rotate the engine a number of times with the ignition switched off, or better still, rotate the engine in the reverse direction by engaging top gear and pushing the car backwards.

Checking the Sparking Plugs. We now revert to our test of one sparking plug. It must be emphasized that the fact that a spark occurs across the points is by no means conclusive proof that the ignition system is above reproach. If the sparking plug insulator is defective, dirty, or damp with condensation, the spark may take place across the points at normal atmospheric pressure, but will probably short-circuit across the internal or external surface of the insulator when the plug is subjected to compression in the cylinder, as explained in Chapter IV.

The end of the plug lead should, therefore, be held or secured

CHART No. 1

DIFFICULT STARTING AT NORMAL TEMPERATURES

- **Engine will not Start** (Starter turns engine at satisfactory speed)
  - Flood carburettor or check that petrol reaches float chamber (see text)
    - Petrol available at carburettor
      - Check spark at plug points (see text)
        - Spark occurs at plug points
          - Inject petrol into cylinders and attempt to start engine
            - Engine fires several times
              - Carburettor troubles (see Chapter V).
              - Air leak in induction system.
            - Engine does not fire
              - Mixture too rich—ventilate engine.
              - Mechanical fault.
              - Incorrect timing.
        - No spark occurs at plug points
          - Spark occurs across $\frac{1}{4}$ in. gap between end of plug lead and cylinder head (see text)
            - Sparking plug faults (see Chapter IV).
            - Spark too weak to fire mixture.
          - No spark from plug lead to cylinder head
            - Spark occurs across $\frac{1}{2}$ in. gap between distributor high tension lead and cylinder head (see text)
              - Faults in distributor cap or H.T. lead, rotor or brush
              - (see Chapter IV).
            - No spark from distributor high tension lead
              - Defective H.T. L.T. leads.
              - Contact breaker faults (see Chapter IV)
              - Faulty Coil.
    - Petrol not reaching carburettor
      - Verify that there is petrol in tank
        - Check delivery of fuel with pipe disconnected from pump outlet union
          - Petrol delivered by pump
            - Carburettor filter choked.
            - Needle valve stuck.
            - Pipe line blocked.
          - No petrol delivered
            - Pump filter choked (see Chapter V).
            - Faulty pump (see Chapter V).
            - Pipe line from tank to pump blocked or leaking air

- **Engine Starts but Stops Immediately** — Check petrol supply (see above)
  - Adequate petrol supply — Check ignition system (see above)
    - Ignition system in good condition — Carburation faults
      - Mixture control incorrectly set.
      - Faulty carburettor needle valve.
      - Sticking carburettor piston.
      - Water in carburettor.
      - Exhaust tail pipe blocked.
    - Ignition system unsatisfactory
      - Loose connexion in low tension or high tension circuit.
      - Contact breaker sticking.
      - Ignition faults (see Chapter IV).
  - Inadequate petrol supply — Check operation of petrol pump
    - Pump operating
      - Filters choked.
      - Air vent hole in petrol tank blocked.
    - Pump not operating
      - Faulty pump (see Chapter V).
      - Intermittent electric disconnexion (electric pump).

- **Starter does not Turn Engine** — Check brilliance of headlights
  - Lights bright
    - Defective starter switch or lead
      - See Chaper XIV.
    - Starter switch and leads in order
      - Wrong grade of engine oil in use.
      - Defective starter (see Chapter XIV).
      - Water pump frozen.
  - Lights dim or fail to light up
    - Battery discharged.
    - Battery connexion or leads defective.

about $\frac{1}{4}$ in. from an earthed bare metal surface while the engine is rotated with the ignition switched on. A healthy, bluish spark should jump the $\frac{1}{4}$ in. gap. If this happens, the sparking plugs must be the cause of the trouble. Otherwise, attention must be turned to the ignition distributor and coil.

Coil Ignition Tests. The next test detailed in the chart will conclusively determine, on coil ignition systems, whether the ignition coil, contact breaker points, condenser and low-tension circuit are in good condition. The test consists of removing the end of the main high tension lead, which runs from the coil to the distributor cap, at the point at which it enters the cap. Take off the cap itself, so that the contact breaker mechanism is exposed. See that the contact breaker points are in contact with one another, if necessary rotating the engine until they close.

This, incidentally, will immediately reveal faults such as a stuck rocker arm which carries the movable point, caused by the fibre of the arm swelling and binding on the pivot pin. It may be necessary to enlarge the hole slightly, although lubrication with graphite from a lead pencil is often sufficient to cure the trouble. A broken rocker-arm return spring will also be obvious on inspection. A get-you-home dodge is to force a wedge of india-rubber, cut from an eraser, or made by rolling up a small section of tyre inner tube, between the rocker arm and the side of the contact breaker case, so that the points are closed. A short length of flexible wire will also be needed to make a connexion between the contact on the rocker arm and the broken portion of the spring, or the pillar to which it is attached.

On the other hand, if the contact breaker appears to be in order, hold the bare end of the high tension lead $\frac{1}{2}$ in. away from an earthed metal surface, and flick apart the contact points. A good spark across the $\frac{1}{2}$ in. gap isolates the trouble in the " spark distributing " components of the distributor. Some likely faults are discussed in Chapter IV. Conversely, lack of a spark indicates coil, contact breaker, or low tension circuit troubles.

When Starter will not Rotate Engine. When the starter motor will not rotate the engine, or will turn it over only sluggishly, it should be a fairly simple matter to trace the trouble on the lines indicated in the chart. At normal

CHART No. 2

COLD-WEATHER STARTING TROUBLES

Difficult starting in cold weather can be caused by any of the faults listed in Chart No. 1, in addition to the points given below, which are peculiar to cold-weather operation.

ENGINE WILL NOT FIRE

- Cranking speed too low — See "STARTER DOES NOT TURN ENGINE," Chart No. 1.
- Cranking speed exceeds 30–60 r.p.m. — Check setting of starting controls (see text). — Controls correctly set — Check petrol level in carburettor, after priming when mechanical pump is fitted
  - Level correct — Check for spark at plugs see Chart No. 1
    - Good Spark — Carburation faults
      - Choked starting jet.
      - Starting device inoperative.
      - Stale petrol in float chamber.
      - Induction manifold too cold—(see text).
    - Weak or no spark — Ignition faults
      - Condensation on Sparking plug insulators—(see text).
      - Condensation on ignition system components.
      - Coil ignition faults—see Chart No. 1.
      - Magneto faults—impulse starter defective.
  - Level too low — Petrol supply system faults—see Chart No. 1 and Chap. V.

ENGINE FIRES ONCE OR TWICE BUT WILL NOT RUN — Examine sparking plug points and insulators

- Condensation on points and insulators
  - Sparking plugs too cold—(see text).
  - Induction system too cold.
  - Excessive flooding of carburettor.
- Insulators clean and dry
  - Mixture too weak.
  - Battery discharged (see text).
  - Engine too stiff due to wrong grade of oil or gummy deposit.
  - Valves sticking (see Chapter III)

temperatures the most likely causes are battery or starter defects, faulty leads, or poor contacts, possibly in the starter switch itself.

In cold weather, too heavy a grade of engine oil will overload the starter motor, while a frozen water pump will effectively lock the engine against rotation—unless the pump spindle shears. This, however, brings us to the question of cold weather starting troubles, when the low atmospheric temperature introduces a new set of problems.

Cold Weather Starting Troubles

Modern engines are generally subjected to exhaustive tests during the initial development period to ensure that they will start easily even at sub-zero temperatures. Provided that carburation, ignition, and compression are satisfactory, therefore, difficult starting should not be experienced if the engine is rotated at an adequate speed. The last proviso often proves the real stumbling block.

The high-speed engine of to-day is to a certain extent a compromise : large inlet valves, smoothly swept induction manifolds, and overlapped valve timing all contribute to a high power output at engine speeds in the neighbourhood of 4,000 r.p.m., but inevitably result in reduced charge velocity at low cranking speeds. This in turn means poor atomization of the particles of petrol which are suspended in the air stream, and a greater likelihood that the fuel will condense on the surfaces of the manifolds before reaching the combustion chambers.

As a result, the average engine must be rotated at between 40-60 r.p.m. to ensure a certain start in cold weather. In mild weather, or in a warm garage, a cranking speed of 100-120 r.p.m. is generally attained if the battery is fully charged and the starter motor in good condition. When the car has stood outside overnight or in an unheated garage, however, the starter is rarely capable of rotating it at 60 r.p.m. initially.

Apart from the low gas speed just referred to, this introduces another drawback on engines fitted with coil ignition systems. At low cranking speeds a very high current is taken by the starting motor, often reaching 400 amps. This means that the battery voltage is reduced to half the normal figure, so that the ignition coil is unable to generate a sufficiently high voltage to provide the hot, efficient spark so necessary under cold-starting conditions. On a magneto the low speed similarly reduces the intensity of the spark, unless an impulse starter is fitted.

Ensuring an Easy Start. Such is the importance of an adequate cranking speed that this is emphasized as the first

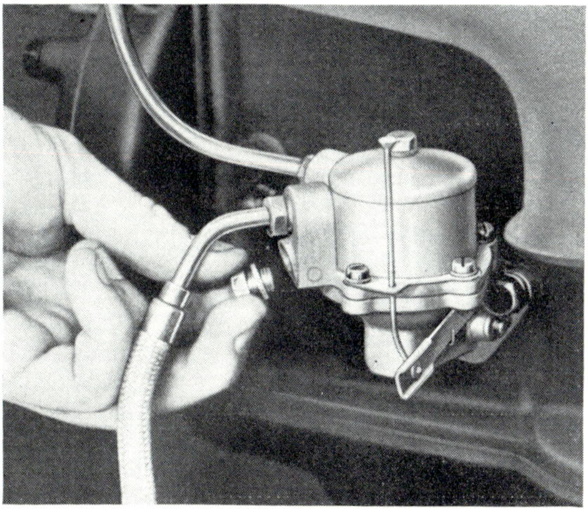

FIG. 5. A MECHANICAL PETROL PUMP. THE PRIMING LEVER, USED WHEN STARTING FROM COLD, AND DRAIN PLUG ARE NOT FITTED TO ALL PUMPS

point to check on Chart No. 2. Sometimes an engine will fire once or twice, or spit back after the starter has been used for several seconds. If the engine is immediately swung sharply by means of the starting handle, an instant start is often obtained : the cylinders are full of combustible mixture, as evidenced by the attempt to fire, and the full battery voltage, when the battery is relieved of the starter load, is sufficient to provide an adequate spark.

Most of the items on the chart are self-explanatory, and provided that the logical sequence of tests is followed it should be possible to track down the cause of the difficult cold starting.

The importance of correctly setting the starting control should not need emphasis. Modern carburettors are fitted with various cold-starting devices which are described in detail in Chapter V. When a self-contained "starting carburettor" is embodied in the main instrument the starting mixture is discharged on the engine side of the throttle, which should, therefore, be fully closed when starting. If the accelerator pedal is depressed, the suction on the starting carburettor is reduced, and the correct mixture will not be obtained.

When an automatically-operated choke or "strangler" is fitted, the mechanical linkage, thermostat, and spring-controlled air valve must be checked over. Often a choke of this type is interconnected with the throttle linkage so that the throttle is automatically opened to the correct extent when the choke is in use: this is another point worth checking.

With some carburettors the dashboard starting control must be pulled fully out, or pressed right home—a half-way position will upset the operation of the starting device. On many modern carburettors and earlier types fitted with simple chokes or stranglers, however, this does not apply. A final point regarding earlier types of carburettor is that often a start in really cold weather can be obtained only by flooding the float chamber; a plunger is generally fitted for this purpose when flooding is necessary.

Evaporation of Petrol. Reference is made in the chart to priming the carburettor when a mechanical pump is fitted. The under-bonnet warmth which exists for a considerable period as the engine cools down causes evaporation of the fuel in the float chamber. An electric pump will beat several times when the ignition is first switched on, but when a mechanical pump is fitted the float chamber will not be refilled until the engine has been rotated a number of times. Since these initial revolutions merely cause a drain on the battery, the use of the priming lever on the pump, when fitted, is a wise precaution.

Unfortunately, it is the lighter fractions in the fuel which evaporate first, and normally it is these more volatile fractions which provide an easy start. Even though the float chamber may be topped-up with fresh petrol by the pump, therefore, the fuel will still contain a large proportion of heavy fractions. This state of affairs will not be remedied until all the original petrol has been drawn from the float chamber and replaced by fresh from the tank; naturally this takes a considerable time, and the battery may be exhausted in turning the engine over meanwhile, or severe condensation may be experienced.

To make sure of a start it is a good plan to empty the float chamber by removing the sediment plug, or by removing the chamber and pouring out the fuel, depending on the design of the carburettor. The bowl can then be refilled by the petrol pump ; better still, fill it from a bottle of cigarette lighter fuel —a useful source of highly volatile petrol.

Apart from carburation, ignition, electrical or mechanical faults, which are fully covered in the appropriate chapters, condensation is one of the most likely troubles. Condensation of moisture on the external surfaces of the sparking plugs, distributor cap, coil, and high tension leads, is a frequent cause of difficult starting. Wipe the surfaces with a clean, dry cloth, or, better still, with a cloth damped with methylated spirit. The spirit (alcohol) will absorb the moisture, and will quickly evaporate, leaving the surfaces clean and dry.

The Application of Heat. To dispel internal condensation heat is the only effective remedy. Condensation of petrol on the walls of the induction manifold has already been referred to. Apart from the effect of a low atmospheric temperature, the induction manifold is still further cooled by the refrigerating effect of the evaporating fuel as it leaves the carburettor. In addition to causing the fuel to recondense, the temperature of the manifold is often lowered to such an extent that the fuel actually freezes on the walls if repeated attempts are made to start the engine—this freezing can take place even when the atmospheric temperature is above freezing point.

Heat, then, is the remedy : it is surprising how little is required to evaporate the condensed petrol. If the radiator has been drained and can be refilled with warm water, starting difficulties usually disappear. If this is impracticable, try the effect of a rag dipped in boiling water, wrung out, and wrapped around the induction manifold. Failing either of these expedients, allow the engine to stand for five minutes after the initial attempts to start. This will allow some of the condensed petrol to re-evaporate, while the battery will also be given a chance to recuperate. A start can often be obtained at the first attempt after a short wait.

Condensation Troubles. Condensation of water or petrol on the electrodes and internal insulators of the sparking plugs is frequently experienced after several attempts to start a cold engine, particularly if the engine has fired once or twice. Every gallon of petrol burnt produces more than a gallon of water, so that the presence of condensed drops of water in the combustion chamber is not surprising. The quickest method of drying the plugs is to remove them, support them in a vertical position in a safe place, and pour petrol into them. If this is lighted and allowed to burn away, the plugs will be both dry and warm for refitting to the engine.

A further point is that some drivers maintain that a good start will be ensured the next day by speeding up the engine, switching off the ignition, and closing the choke or keeping the starting control in operation until the engine stops. The theory is that the rich mixture introduced into the cylinders will provide a suitable starting mixture when the car is next used. Whether the method is effective depends to a great extent on the characteristics of the engine. In any case, the fuel which is deposited on the cylinder walls as the engine cools down is liable to dilute or wash away the oil film, leading to rapid cylinder-wear, while condensation on the sparking plug points and insulators may cause difficult starting. When condensation trouble is experienced the alternative scheme of revving up, switching off, and holding the throttle wide open to introduce as dry a mixture as possible, is often effective.

Misfiring and Loss of Power

Some causes of misfiring and loss of power are apt to be extremely elusive. By following a logical process of diagnosis as set out in Chart No. 3, however, it should be possible to track down the trouble without undue delay. Nevertheless, one or two points deserve amplification as an aid to diagnosis.

Usually, the loss of power which occurs when a car has covered a considerable mileage is the result of an accumulation of minor faults rather than a major defect ; as a result the process is so gradual that the average driver is apt to be unaware that the car's performance has deteriorated.

That some loss of power is experienced when even 10,000 miles have been covered is not surprising when it is realized that, during this period, on an average car, each piston travels about 7,000 miles up and down the cylinder walls. Each sparking plug fires, and each valve opens and closes, more than 18,000,000 times. The ignition coil must provide more than 72,000,000 sparks, the contact breaker points opening and closing a similar number of times.

The carburettor must mix about 400 gallons of petrol with 700,000 cubic feet of air, and the fuel pump must operate about 20,000,000 times to provide sufficient petrol. Meanwhile the battery has received considerably more than 2,000 horse-power in the form of electrical energy from the dynamo.

Using a Vacuum Gauge. One of the most useful instruments in tracing loss of power and misfiring is a vacuum gauge, which, as mentioned in Chapter I, forms part of most proprietary engine test sets. Any vacuum gauge, preferably calibrated in inches of mercury, however, can be used to diagnose a number of troubles.

The gauge should be connected to the induction manifold close to the carburettor. When a union is provided for a suction windscreen wiper, vacuum-operated ignition control, or vacuum-operated chassis lubrication pump, connexion of the gauge pipe line is a simple matter. Otherwise it will be necessary to drill and tap the manifold to take a suitable union, although this can be obviated by the use of an adaptor, supplied with some gauges, which is fitted between the carburettor and inlet manifold flanges.

With the engine warmed-up and running at a speed slightly faster than the normal idling setting, the following readings should be obtained on the gauge, depending on the altitude above sea-level at which the test is made and the particular engine—

Altitude	Inches of Mercury
Sea-level	18 – 20
1,000 ft	17·5 – 19·1
2,000 ,,	16·8 – 18·5
3,000 ,,	16 – 17·8
4,000 ,,	15·2 – 17
5,000 ,,	14·6 – 16
6,000 ,,	14 – 15·4
7,000 ,,	13·2 – 14·8
8,000 ,,	12·6 – 14
9,000 ,,	12 – 13·2
10,000 ,,	11·2 – 12·6

If we assume a normal reading of 18–20 inches at sea-level, a lower reading can be due to one or more of the following

CHART No. 3

MISFIRING AND LOSS OF POWER

Many of the causes of misfiring and loss of power can be logically diagnosed by following the appropriate sections of Chart No. 1. In the chart below, however, the most likely causes of misfiring and loss of power, as opposed to difficult starting, have been segregated.

- **Engine Misfires or Lacks Power**
  - At high speed — Check spark as in Chart No. 1
    - Satisfactory spark
      - Ignition faults (see Chap. IV).
        - Defective plugs.
        - Wrong grade of plug.
        - Insufficient ignition advance.
        - Automatic advance mechanism sticking.
      - Carburation faults (see Chap. V).
        - Main or power jet choked.
        - Fuel starvation.
        - Throttle not opening fully.
      - Mechanical faults (see Chap. III).
        - Incorrect valve clearance.
        - Weak valve springs.
        - Sticking valves.
        - Incorrect valve timing.
        - Partial seizure.
    - Unsatisfactory spark — Ignition faults — overhaul system as described in Chapter IV.
  - At low speed
    - Spitting-back in carburettor — Check fuel supply as in Chart No. 1
      - Supply adequate
        - Choked jet.
        - Water in carburettor.
        - Sticking valve.
        - Pre-ignition.
        - Incorrect valve timing.
      - Supply inadequate — Fuel feed troubles (see Chart No. 1 and Chap. V).
    - No spitting-back in carburettor — Check spark and ignition system as in Chart No. 1 — Satisfactory spark — Check for pinking at full throttle opening
      - Pinking
        - Ignition too far advanced.
        - Advance mechanism stuck.
        - Wrong grade of plug.
        - Incandescent particles in combustion chamber
      - No Pinking
        - Poor compression due to leaking valves, piston rings, etc. (see Chart No. 4).
        - Ignition too far retarded.
        - Exhaust system choked.
        - Incorrect valve timing.
        - Partial seizure.

- Explosions in silencer
  - On over-run with throttle closed, at low, medium or high speeds
    - Check exhaust system for air leaks
      - No air leaks
        - **Carburettor idling, mixture too weak.**
        - Intermittent short circuit in ignition system.
        - Exhaust tail pipe too short.
  - When engine is pulling
    - Check ignition system as in Chart No. 1
      - Ignition system satisfactory
        - Sticking exhaust valve.
        - Incorrect valve timing.

- ENGINE CONTINUES TO RUN WHEN SWITCHED OFF
  - Verify that correct grade of sparking plug is fitted
    - If correct grade of plug fitted
      - Engine overheating (see below)
      - Engine not overheating
        - Exhaust valve overheating or badly burnt.
        - Incandescent particles in combustion chamber.
        - Fault is a peculiarity of some engines and is then extremely difficult to cure.

- ENGINE OVERHEATS
  - If ignition is not too far retarded
    - Check mixture strength
      - Mixture correct
        - Mechanical faults (see Chapter III)
          - Pistons and bearings tight.
          - Engine needs decarbonizing.
          - Lack of oil.
          - Brakes binding (see Chapter X).
        - Cooling system faults (see Chapter VI)
          - Shortage of water.
          - Water passages in radiator and cylinder block choked.
          - Fan or water pump inoperative or inefficient.
          - Thermostat defective.
      - Mixture incorrect
        - Carburettor faults (see Chapter V).
        - Fuel feed faults (see Chapter V).
  - Ignition too far retarded
    - Check distributor and timing
      - Contact breaker gap too small (see Chap. IV).
      - Automatic advance mechanism stuck (see Chapter IV).
      - Wrong timing (see Chapters III and IV)

STEADY NEEDLE

18°—21°

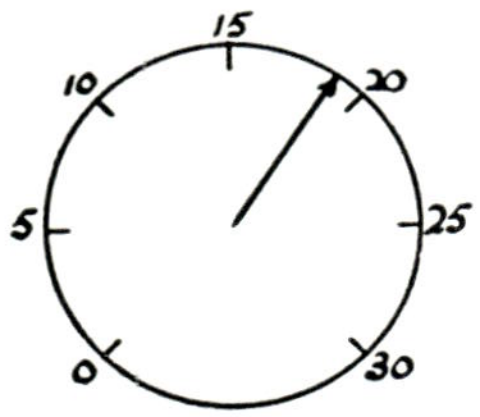

Normal Vacuum prior to REDEX. Rising 10% or more after REDEX.

STARTER VACUUM TEST

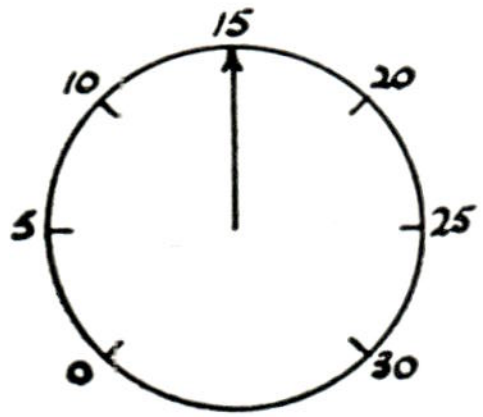

Engine warm. Butterfly completely closed. Operate starter with ignition off. Normal engine should give 15° or over. Low reading indicates worn butterfly, induction leaks, perforated hot-spot.

GENERAL CHECK ON ENGINE CONDITION

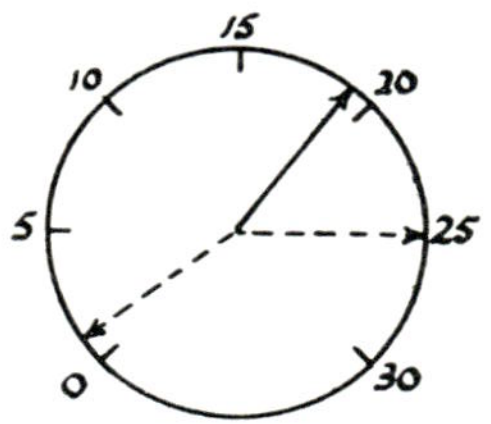

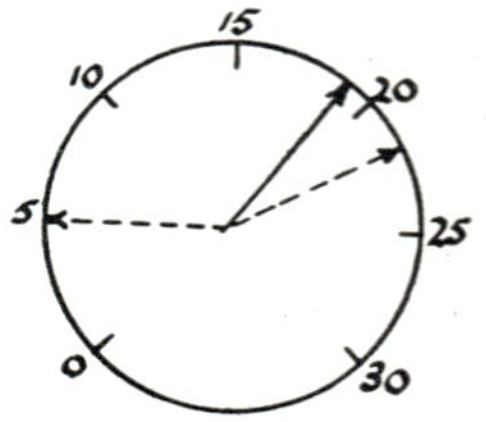

OPEN AND CLOSE THROTTLE QUICKLY

Needle falls sharply to zero and returns quickly to about 25°, then steadies to normal. Indicates good engine.

Needle falls slowly to about 5°, then drifts back to normal. Engine in poor condition generally. Very slow return of needle indicates choked exhaust.

CHECKING AUTO-ADVANCE

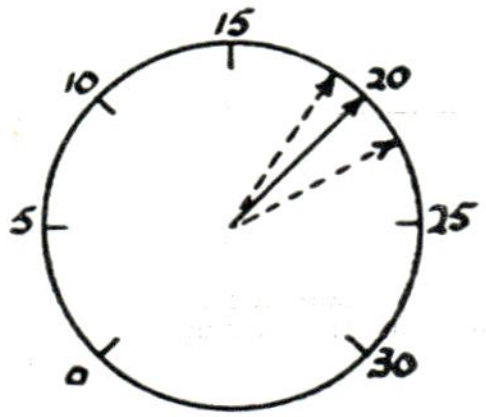

Increase engine speed gradually to 1200 rpm or more. If needle drops $\frac{1}{2}$-1° then rises 1-1$\frac{1}{2}$° auto-advance O.K.

FIG. 6. PRELIMINARY CHECKS ON ENGINE CONDITION, WITH THE AID OF A VACUUM GAUGE

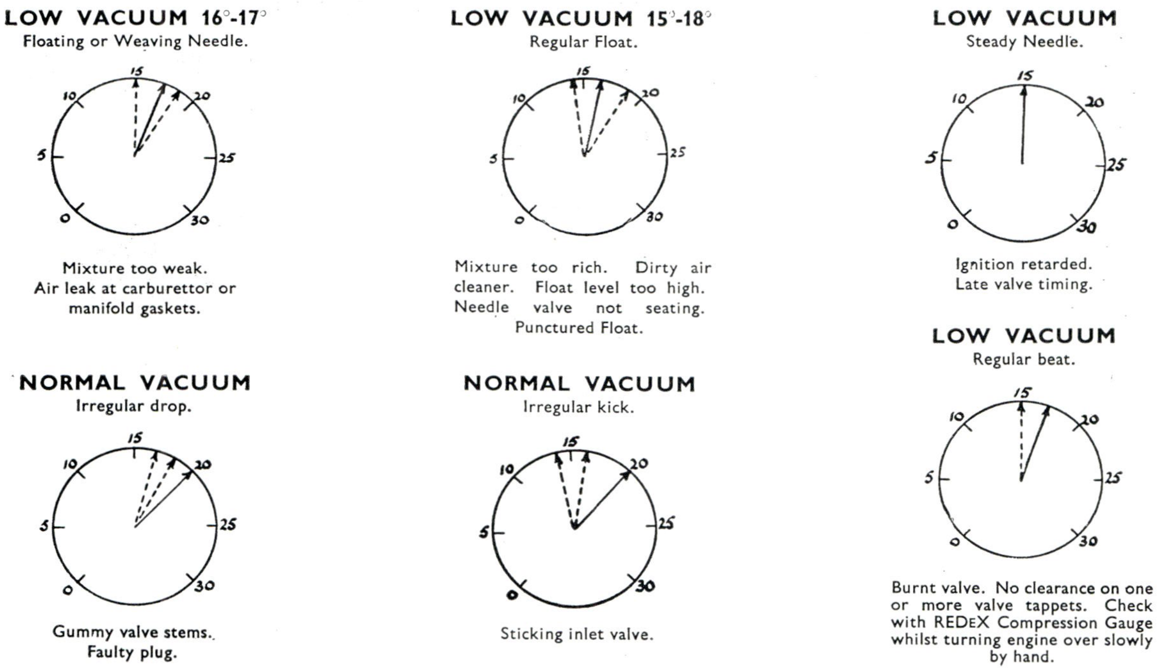

FIG. 7. VACUUM GAUGE READINGS AS AN AID TO FAULT DIAGNOSIS

SLIGHTLY LOW VACUUM

Needle tremor.

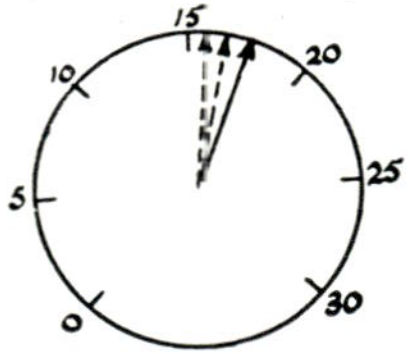

Tappets out of adjustment.

CHECKING VALVE SPRINGS

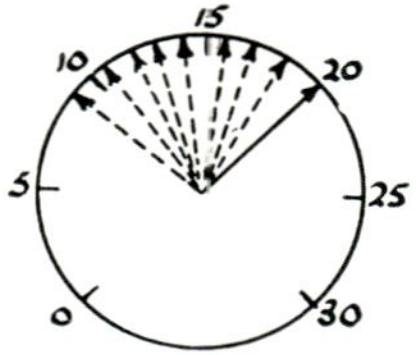

Increase engine speed gradually. Needle falls and vibration commences increasing with engine speed. Indicates weak valve springs.

NORMAL VACUUM

Irregular kick.

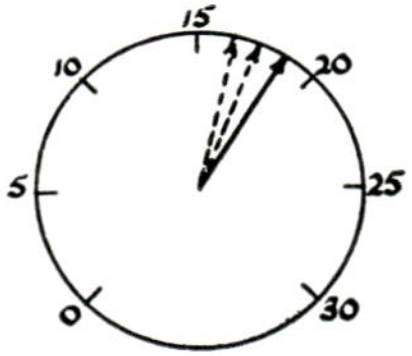

Plug points too close. Plugs fouled. Contact points pitted, burnt, or out of adjustment. Auto-advance springs loose.

LOW VACUUM

Very unsteady needle.

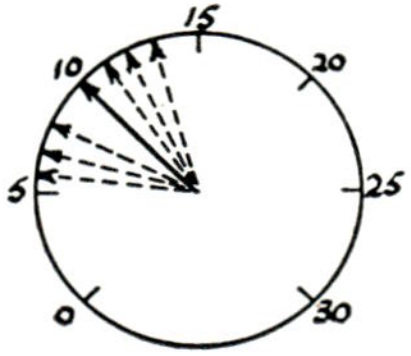

Hot-spot plate or heat riser burnt through.

Starter vacuum test will assist diagnosis.

CHECKING PLUGS

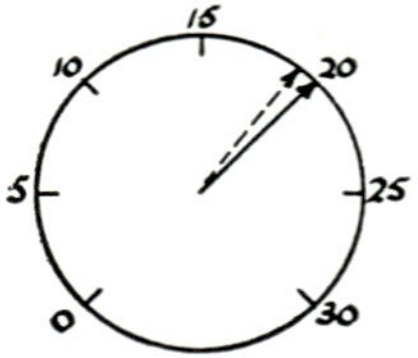

Short each plug in turn. If good, needle drops:—$1\frac{1}{2}$-2° for 4 cylinder; $\frac{3}{4}$-$1\frac{1}{2}$° for 6 cyl.; $\frac{1}{4}$-$\frac{3}{4}$° for 8 cyl. If still no drop after plugs tested cylinder is faulty, i.e. burnt out exhaust valve. Confirm with Compression Gauge

HIGH VACUUM

Irregular kick.

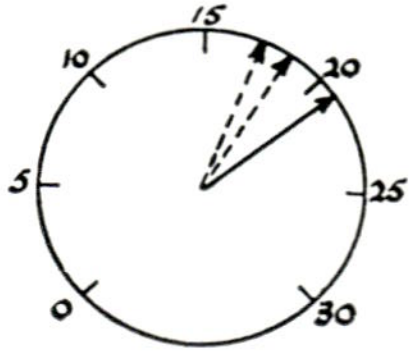

Over-advanced. Loose timing chain. Faulty condenser.

FIG. 8. MORE VACUUM GAUGE INDICATIONS OF COMMON FAULTS

faults: loss of compression; incorrect ignition-timing; incorrect valve-timing; incorrect carburettor-setting (the needle will also "float"). The various faults are dealt with in greater detail in the appropriate chapters.

Checking the Ignition System. Apart from the more detailed tests described in Chapter IV, the vacuum gauge will give a fairly good indication of the efficiency of the sparking plugs. By short-circuiting each plug in turn, a defective plug will be revealed as the one which causes the least drop in the vacuum gauge reading when short-circuited. It is possible, however, that faulty valves or piston rings in that cylinder may be the culprits, a fact which can be verified by cleaning and testing the plug, or by fitting a replacement known to be in good condition, and repeating the test. If the vacuum gauge reading does not drop to as great an extent as when the other plugs are short-circuited, the trouble lies in the pistons, valves or gasket.

To check the action of the automatic advance mechanism, speed up the engine progressively from idling speed ; the gauge reading should suddenly advance by from one-half to two inches as the control comes into action.

Valve and Tappet Checks. Tremor of the needle indicates incorrect tappet adjustment. On four-cylinder engines, however, some tremor is generally experienced at low speeds due to the pulsations in the induction manifold. Usually, proprietary gauges have a damping control which should be screwed in to give a steady reading. Alternatively, the rubber connecting tube may be compressed to give the same effect. Six-, eight-, and twelve-cylinder engines normally give a steady reading, and there is no necessity for damping out pulsations in the gauge line.

A sticking valve is indicated by sudden drops in the reading as the engine speed is slowly increased. Weak valve springs and valve bounce are revealed by erratic readings at or near maximum engine speed. Some notes on diagnosing valve troubles are given in Chapter III.

Cylinder and Ring Wear. When the throttle is suddenly closed with the engine running at high speed the gauge reading should momentarily increase to 5 inches above the normal idling reading. Failure to reach the increased reading indicates worn cylinder bores, worn, broken, or gummed piston rings. Further diagnosis of these faults is covered in Chart No. 4.

Choked Exhaust System. A choked or partially restricted exhaust system is a source of power loss which is often overlooked. Owing to the overlapped valve timing used on most modern engines, the exhaust valve remains open after the piston has begun to descend on the inlet stroke, in order to induce a process known to technicians as " inertia scavenging." The inertia of the swiftly moving exhaust gases in the exhaust pipe and manifold creates a suction which tends to draw the remaining spent gases out of the combustion chamber. Naturally, if the velocity of the exhaust gases is reduced, or if actual back pressure is set up because of accumulations of carbon in the exhaust pipe and silencer, a dislodged silencer baffler plate, or similar causes, the exhaust gases will not be effectively scavenged from the combustion chamber, and the incoming mixture will be diluted.

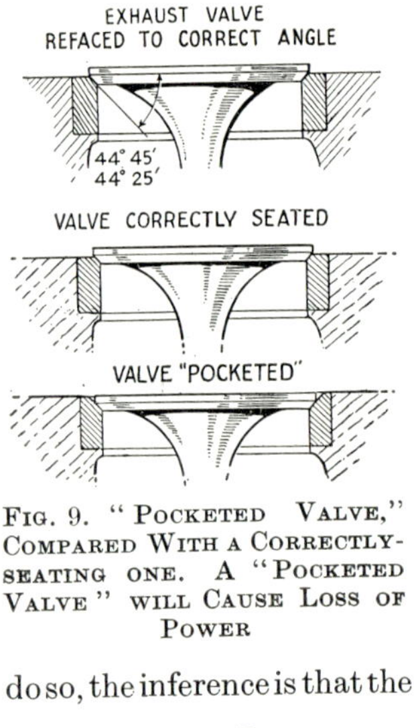

Fig. 9. " Pocketed Valve," Compared With a Correctly-seating one. A "Pocketed Valve" will Cause Loss of Power

The efficiency of the exhaust system should be checked, therefore, by pressing a folded rag against the end of the tail pipe while the engine is running at a moderately high speed. The needle of the vacuum gauge should then drop back by several divisions. If it does not do so, the inference is that the exhaust system is partially choked.

Detonation and Pre-ignition

Most of the points outlined in Chart No. 3 covering detonation or " pinking " under load, pre-ignition and auto-ignition or " running-on " when the engine is switched off, are self-explanatory.

It is as well, however, to differentiate between detonation and pre-ignition. Detonation is the term used to describe the spontaneous explosion of the remaining charge in the combustion chamber when only a portion of it has burnt progressively. Normally the flame should spread outwards through the charge from the sparking plug at a rate which is controlled by the design of the combustion chamber, so that the piston receives a progressive downward push. If, however, the burning charge raises the pressure in the combustion chamber to too high a degree, the fuel on the side of the combustion chamber remote from the advancing flame front will explode spontaneously. The piston thus receives a sharp blow instead of a progressive push, and a characteristic metallic knock or " pinking " is audible. If the initial compression of the charge is high, the fuel will be more liable to detonate, unless a petrol with a high anti-knock or octane value is used, Thus sports car engines require a higher octane value fuel than do " touring " engines.

Pre-ignition and auto-ignition differ from detonation in that the incoming charge is ignited by an incandescent particle or surface before the sparking plug fires. Thus the piston receives its thrust too early, and rough running and knocking are experienced.

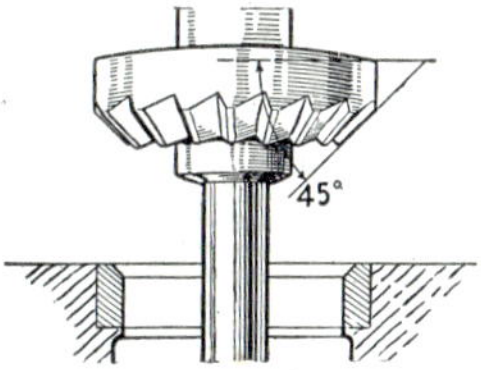

FIG. 10. BADLY PITTED SEATINGS SHOULD BE RECONDITIONED BY THE USE OF A VALVE RESEATING CUTTER

Assuming that the fuel is suited to the compression ratio of the engine, the detonation and pre-ignition can be due only to local overheating at some point in the combustion chamber. For instance, the sparking plugs may not be able to withstand the heat developed at full throttle or high speed—see Chapter IV—or the combustion chamber and piston crown may be coated with a heavy layer of carbon which not only raises the compression ratio, but may also cause pre-ignition if particles of carbon become incandescent. Similarly, a fragment of copper from the gasket projecting into the combustion chamber will become white-hot and fire the charge prematurely. Cooling system faults may cause the engine to overheat, and local boiling may take place around the combustion chamber, forming steam pockets which lead to a rapid rise in temperature at points such as the exhaust valve seatings or sparking plug bosses.

Similarly, running-on when the ignition has been switched off is due to some part, such as the sparking plug points or the exhaust valve, becoming sufficiently hot to fire the charge. Some engines, usually of the sports type, are very prone to this fault, and it is often difficult to cure it without upsetting the

FIG. 11. A HEAVY CARBON DEPOSIT WILL CAUSE DETONATION AND PRE-IGNITION

The piston illustrated also shows evidence of severe scoring and partial seizure, while the rings are scored and stuck in their grooves

performance of the engine under normal conditions. Generally the best compromise is to allow the engine to idle for a few moments before switching-off.

Explosions in Silencer. A word of explanation is perhaps required regarding the correct diagnosis of the cause of explosions in the silencer. Obviously, explosions can be caused only by combustible mixture reaching the silencer, where it is fired by an incandescent particle of carbon or by a succeeding exhaust flame. A sticking exhaust valve may allow mixture to escape into the exhaust system, or a faulty sparking plug, or an intermittent short-circuit in the ignition system, may allow unburnt mixture to be exhausted from the cylinder.

In the majority of cases, however, the cause of explosions in the silencer, which are generally experienced when the throttle is closed at moderately high engine speeds, is nothing more serious than an air leak in the exhaust system.

The explanation is as follows : when the throttle is closed while the engine is running at high speed, the air supply to the

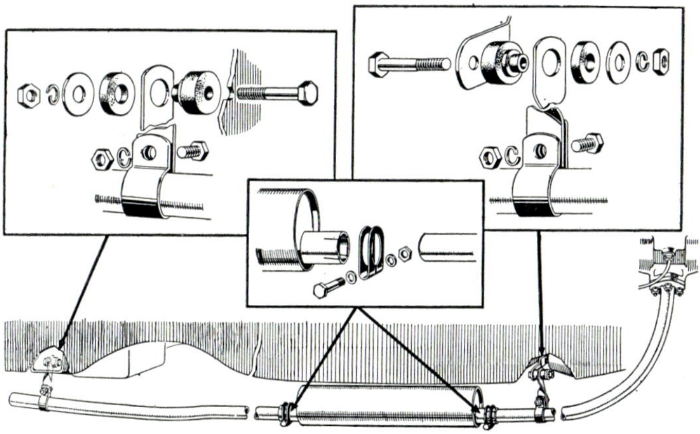

FIG. 12. THE EXHAUST SYSTEM SHOULD BE CHECKED FOR LEAKAGE AT THE JOINTS, A CHOKED SILENCER OR TAIL PIPE, OR DISPLACED SILENCER BAFFLES

engine is practically cut off, and partially burnt mixture passes into the exhaust system. This mixture is capable of further combustion if additional oxygen is available ; an air leak in the system provides the necessary oxygen, so that explosions occur.

The fact that the explosions can often be cured by slightly enriching the slow-running mixture is apt to be misleading. The result, of course, is still further to enrich the partially burnt mixture, so that more oxygen is required before it will burn in the silencer. Unless the air leak is fairly extensive, therefore, combustion does not take place in the silencer. The " cure " is not economic.

The proper procedure is to check carefully the exhaust manifold flange nuts, washers, and all joints in the exhaust system, not forgetting the possibility that, owing to rust, pin-holes may have formed in the exhaust pipe and silencer.

Engine Tune-up. With the aid of vacuum and compression gauges and ignition testing equipment an engine can be checked systematically, followed by correct adjustment, resulting in a substantial gain in power, liveliness, and economy without the necessity for any major operations.

The tune-up should be carried out in six successive steps, following the sequence given below, which experience has proved to give the best results.

1. Remove the sparking plugs, check that they are of the correct type for the engine, clean, reset the gaps, and test as described in Chapter IV.

2. Replace a sparking plug in one cylinder at a time and check the compression of each cylinder in turn ; or, preferably, use a compression gauge as described in Chapter III.

Failing the use of a compression gauge, the compression must be judged by the " bounce " felt on the starting handle at top dead centre on the compression stroke. If compression is not good and reasonably even on all cylinders, it will not be possible to tune the engine very effectively.

3. Remove the ignition distributor cap and check it for cracks, " tracking " between the electrodes, burnt electrodes, and similar faults. Check the action of the centrifugal and suction advance mechanisms. Examine the distributor points and if necessary adjust the gap. Check all high and low tension ignition leads. Check the ignition timing. Chapter IV explains the foregoing procedure in some detail.

4. Check the valve clearances, as described in Chapter III, remembering that the clearances should be measured either with the engine cold or fully warmed-up, according to the manufacturer's instructions.

5. Check the carburettor and fuel pump : see Chapter V. Examine, and if necessary clean the air filter. Clean the petrol filters in the pump and carburettor. With the aid of a vacuum gauge, adjust the idling mixture to give the highest possible steady reading. Check the operation of the exhaust manifold heat valve, if fitted.

6. Check the water level in the radiator and, if necessary, tighten all hose clips. Check the tension of the fan belt. Warm-up the engine and check the action of the thermostat.

CHAPTER III

ENGINE: MECHANICAL FAULTS

THE mechanical engine faults covered in this chapter are, of course, closely associated with the operational faults dealt with in Chapter II, since almost any mechanical fault will affect starting, running, or power output in one way or another. Charts Nos. 4 and 5 cover most of the faults likely to be experienced. Again, a logical process of elimination will avoid waste of time, while the use of certain instruments, such as stethoscopes, compression gauges and neon timing lights, will still further cut down the time spent in tracing a particular trouble.

TRACING ENGINE NOISES

Many mechanical faults are accompanied by characteristic noises ; the appearance of an unusual knock, tap or squeak, in fact, may be the first symptom that anything is amiss. Although the section " Engines Noises " in Chart No. 4 is more or less self-explanatory, one or two additional notes may be useful as an aid to fault-tracing.

Identifying Engine Noises. It is not always easy to identify a particular noise owing to the fact that sounds caused by different faults are often somewhat similar in character. Noises are transmitted so readily by metal components, moreover, that the source may be difficult to detect.

Two aids to accurate diagnosis can be employed. The first is a stethoscope ; the second, experience. Of the two, the latter is the more valuable, since a skilled mechanic can often diagnose a noise immediately from past experience, where the novice has to spend a considerable time in locating the trouble. Nevertheless, some form of stethoscope is a useful accessory. A simple version comprises a wooden rod, fitted with a shaped ear-pad, which gives adequate results. Alternatively various types of commercial instrument are available, which are more sensitive. A large screwdriver with a long shank will serve in an emergency, provided that the shank and blade

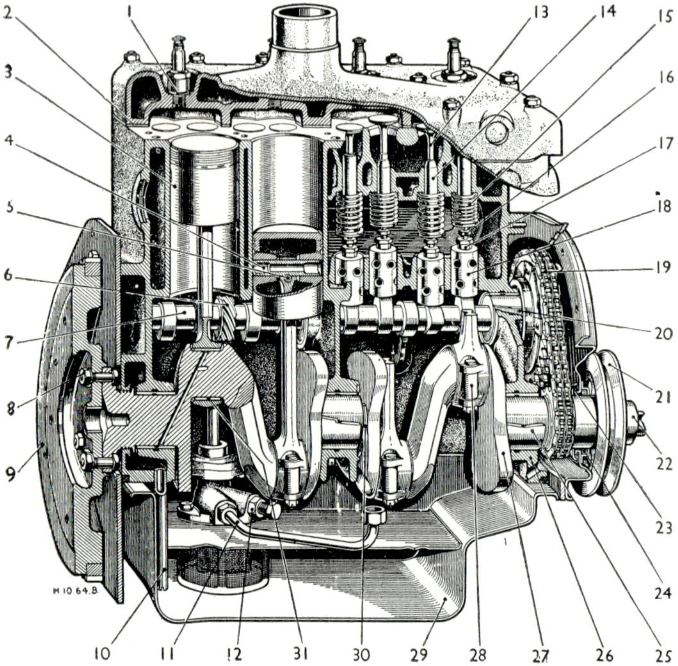

FIG. 13. MECHANICAL FAULTS (SIDE-VALVE ENGINE). A TYPICAL OVERHEAD VALVE GEAR IS SHOWN IN FIG. 14

1. Defective sparking plug and leaking washer
2. Leakage past cylinder head and gasket; loose or unevenly-tightened cylinder head nuts
3. Piston skirt worn or scored; rings worn, broken or stuck in grooves
4. Worn gudgeon pin and bushes
5. Loose gudgeon pin clamping bolt
6. Worn or chipped teeth on oil pump
7. Camshaft bent; cams worn; bushes worn
8. Flywheel retaining bolts loose
9. Flywheel out of balance. Starter ring gear worn
10. Oil return pipe choked
11. Oil pump delivery pipe choked; unions loose
12. Oil pressure relief valve incorrectly adjusted; valve faces dirty or ridged; spring weak
13. Valve stem gummy; worn
14. Valve guide worn
15. Weak or broken valve spring
16. Tappet adjustment incorrect; depression worn on tappet screw
17. Screw thread in tappet stripped
18. Tappet and guide worn
19. Timing chain and sprockets worn; chain tensioner ineffective or incorrectly adjusted
20. Worn camshaft front bearing
21. Crankshaft pulley loose
22. Worn starter dog
23. Bent oil slinger
24. Worn oil retainer
25. Front crankshaft bearing worn, causing oil leakage
26. Bearing cap loose
27. Crankshaft twisted; journals worn
28. Connecting rod bearing worn; bolts loose
29. Oil sump coated with sludge; filter dirty; leakage at joints
30. Incorrect end float on crankshaft
31. Bearing oilway choked

are kept well away from the vicinity of the sparking plug terminals.

Judging the Frequency of a Knock. A knock or tap in an engine can often be localized immediately in a component or group of components by its frequency. If it occurs regularly at approximately engine speed, for instance, it can be attributed to components that rotate or reciprocate at crankshaft speed : in other words, the pistons, connecting rods, flywheel, clutch, and, of course, the crankshaft itself.

If the noise appears to have a frequency of half the engine speed, components driven from the timing chain or gears must be suspect : the camshaft, valves, tappets, rockers, ignition distributor, petrol pump, and oil pump may be at fault.

The Effect of Engine Temperature. Whether the noise is present only when the engine is cold, or becomes less prominent as it warms up, is also a useful pointer. One of the first sounds to be detected in an engine, for instance, is " piston slap." The mileage at which this first becomes noticeable varies widely, depending on the type of engine and the conditions under which it has been used. As the cylinders, piston rings, and piston skirt become worn, the excessive clearance between the piston and the cylinder wall results in the thrust face of the piston striking the cylinder with an audible tap at the beginning of the power stroke.

Since, as the engine warms up, an aluminium piston expands to a considerably greater extent than does the cylinder, the excessive clearance may disappear when the engine reaches its normal running temperature, or the " slap " may be very much diminished.

Modern designs of piston, with oval-ground skirts or special struts that control expansion, allow much closer tolerances to be used during initial fitting of the pistons in the bores, so that the onset of " piston slap " is delayed.

Removing the Load from a Piston. As an aid to diagnosis, each sparking plug should be short-circuited in turn. As the piston is relieved of the explosion pressure in that cylinder, the tap will disappear or become much less noticeable. Short-circuiting the plug will also eliminate the lighter tapping or clicking caused by a worn gudgeon pin and little-end bearing, and usually will cut out the characteristic metallic clatter of a badly worn or " run " connecting rod big-end bearing.

CHART No. 4

ENGINE: MECHANICAL FAULTS

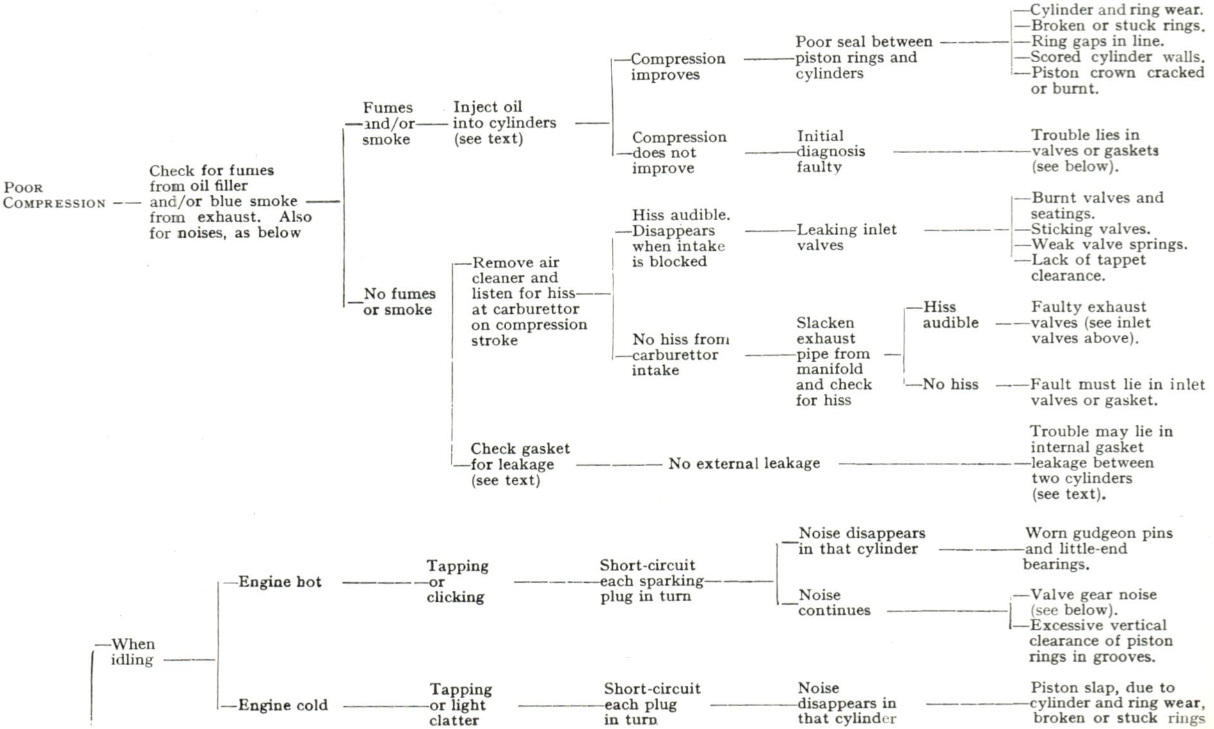

ENGINE NOISES

- **When running light**
  - **Clatter at approx. 2/3 max. engine speed**
    - Short-circuit plug in each cylinder in turn
      - Noise disappears
        - Big end bearing **wear** or run bearing.
- **When under load**
  - High-pitched metallic tapping or " pinking
    - Detonation
      - Engine needs decarbonizing.
      - Ignition too far advanced.
      - Unsuitable fuel.
      - Wrong grade of sparking plugs.
      - Engine overheating.
  - Rumbling or thumping
    - Check oil pressure: if low or leakage is visible at front or rear of crankcase
      - Low pressure leakage.
        - Main bearing wear.
      - Oil pressure normal and no leakage
        - Flywheel loose on mounting.
- **Under all conditions**
  - **Tapping or clicking**
    - Check tappet clearance
      - If correct
        - Worn tappets and guides. (s.v. and push-rod o.h.v. engines).
        - Worn rockers (o.h.v. engines) and shaft.
        - Bent push-rod.
        - Defective silencing springs.
        - Soft spot on tappet or rocker.
        - Worn timing or distributor driving gears.
        - Fan blades striking radiator or hose.
  - Rumbling noise
    - Check for main bearing wear
      - If no wear apparent
        - Worn dynamotor chain.
        - Worn timing gears (but usually a clatter).
        - Slack timing chain (noise generally high pitched).
  - **Lashing, rattling and grinding noises**
    - Check timing chain adjustment
      - If correct or no adjustment provided
        - Worn timing chain and sprockets.
        - Defective tensioner.
  - **Squeaks and whistles**
    - Unlubricated dynamo or fan bearing.
    - Hard or glazed dynamo brushes.
    - Slipping fan belt.
    - Defective water pump bearing.
    - **Unlubricated distributor shaft** bearing.

Little-end bearing noise is generally light in sound, and often disappears if the engine is speeded up, whereas big-end knock is very much louder and is most pronounced when the throttle is partly closed at fairly high engine speeds on the road, or when the engine is speeded up in neutral.

Reducing Valve Gear Noises. The valve gear is often responsible for a variety of tapping and clicking sounds. Even when the tappet clearance is correctly adjusted, and such faults as exist have been remedied as described later in this chapter, valve noise sometimes persists. On push-rod-operated overhead valve engines this may be due to the inertia of the push-rods, which are apt to set up a rattle or clatter at high engine speeds. The remedy is to fit a return spring to the push-rod end of each rocker so that the adjusting screw is kept in contact with the push-rod, and the base of the push-rod with the tappet. An alternative scheme, which is often successful, is to link together pairs of push-rods with light coil springs.

Very often valve noises are amplified by the valve cover, which acts as a sounding board. Among effective remedies is the fitting of cork, oil-resisting rubber, or fibre washers under the heads of the securing nuts, so that the cover is isolated from the valve chest or cylinder head, as the case may be. It is a good plan to line the inside of the cover with felt, secured to the metal with gasket cement. Lightly brush over the exposed surface of the felt with cement also, to prevent any hairs becoming detached and finding their way into the oiling system.

Tracing Compression Leakage

Owing to the fact that on modern cars the starting handle is seldom used—on one or two it is not provided—the compression of the individual cylinders is seldom checked. As a result, leakage past one valve may take place for a considerable time undetected, with the result that the faces of the valve and seating become burnt and distorted. The trouble taken to test the compression from time to time, therefore, will be well repaid.

The test should always be carried out with the sparking plugs removed, except the plug fitted to the cylinder that is

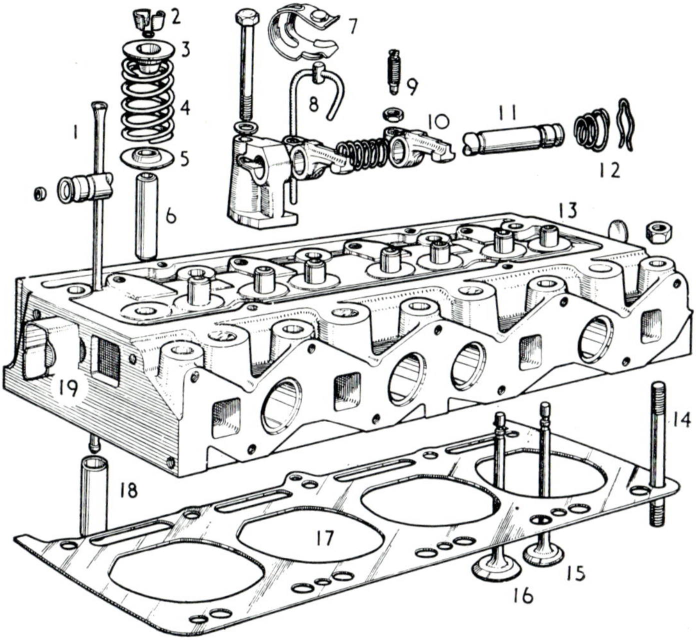

FIG. 14. CYLINDER HEAD AND VALVE GEAR OF TYPICAL OVERHEAD-VALVE ENGINE, SHOWING POSSIBLE FAULTS

1. Bent push-rod
2. Displaced or broken cotters
3. Damaged valve spring cup
4. Weak or broken valve spring
5. Missing valve spring locator
6. Worn valve guide
7. Lubrication pipe bracket damaged
8. Choked lubrication pipe
9. Incorrectly adjusted or worn tappet screw
10. Worn rocker
11. Worn rocker shaft
12. Weak silencing springs
13. Cracked cylinder head
14. Stripped or broken cylinder-head stud
15, 16. Burnt or bent valves
17. Defective cylinder-head gasket
18. Worn tappet
19. Choked or corroded water-distribution tube

being tested. This allows the engine to be rotated more quickly and eliminates any likelihood of confusion between various cylinders. On six-, eight-, and twelve-cylinder engines, in fact, the individual compression can be judged only in this manner.

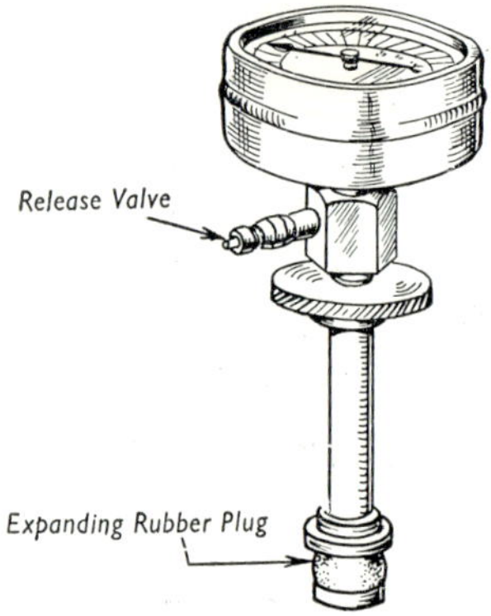

Fig. 15. A Compression Gauge with an Expanding Rubber Connector

Using a Compression Gauge. For accurate tests, a compression gauge is essential. A number of proprietary gauges are available, fitted either with conical rubber connectors or expanding rubber rings to ensure an air-tight seal in the sparking plug hole. An enterprising mechanic, however, can improvise a gauge by soldering a tyre pressure gauge into the body of a sparking plug from which the central electrode has been removed.

The engine should be thoroughly warmed-up, and the sparking plugs removed. The throttle should be kept wide open, and the engine, preferably, rotated by the starting motor, thus ensuring a constant cranking speed. The pressure recorded by the gauge should then be compared with the normal compression for the engine, which is calculated as follows—

Normal compression = (atmospheric pressure × compression ratio) + atmospheric pressure.

The average atmospheric pressure may be determined from the following table—

Altitude	Atmospheric Pressure
Sea-level	14·7 lb per sq in.
1,000 ft	14·1 „ „ „ „
2,000 „	13·6 „ „ „ „
3,000 „	13·1 „ „ „ „
4,000 „	12·5 „ „ „ „
5,000 „	12·3 „ „ „ „
6,000 „	11·6 „ „ „ „
7,000 „	11·2 „ „ „ „
8,000 „	10·8 „ „ „ „

Thus at sea-level the normal compression on an engine having a compression ratio of 6 : 1 would be : $(14{\cdot}7 \times 6) + 14{\cdot}7$, or roughly 103 lb per sq. in.

If a pressure is recorded which is about 10 per cent higher than the normal compression for the engine, this can be considered satisfactory, since compression of the air in the cylinder raises its temperature, which in turn raises its pressure above the figure obtained from the simple calculation given above. As the heat lost to the cylinder wall, piston crown and combustion chamber is negligible, the engineer terms this *adiabatic* compression.

Causes of Low Readings. If all cylinders show a low reading the cause may be cylinder- and ring-wear or badly seating valves. If, on the other hand, only one cylinder shows an abnormally low reading, a sticking valve, broken valve-spring, or fractured or sticking piston rings may be to blame. Nor can the possibility of a scored cylinder wall be ruled out. When two adjacent cylinders show an approximately equal pressure drop, the most likely fault is leakage past the section of gasket which separates them. These narrow strips of gasket often give trouble.

It is a fairly simple matter to find out whether the cylinders and piston rings are at fault by injecting into each cylinder a small quantity of engine oil and repeating the test. If the compression improves, it is likely that leakage is taking place past the piston rings. If, however, too great a quantity of oil is injected, or if it is allowed to flow on to the valves on side-valve engines, any faulty valves may also be temporarily sealed, thus giving a false impression.

Signs of Gasket Leakage. Methods of diagnosing leakage

past the valves are given in Chart No. 4. If the valves are not at fault, the gasket should be checked by applying oil to the external edges and around the cylinder head nuts, when bubbles and froth will reveal any leakage.

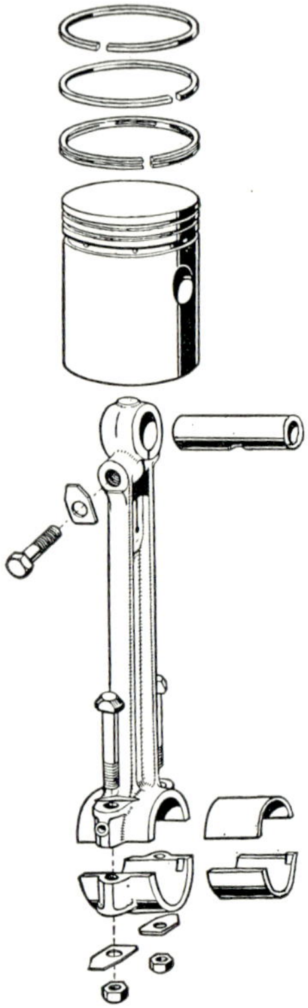

Fig. 16. A Typical Piston, Ring, Connecting Rod and Bearing Assembly

This test is best made with the engine running, since the explosion pressure is very much higher than the compression pressure. Leakage past the gasket into the water passages will be revealed by bubbles in the radiator header tank when the engine is running, probably accompanied by oily scum on the surface of the water.

If the leakage is at all serious, water will usually be found on the internal insulator and points of the sparking plug of the affected cylinder when the engine has been allowed to cool down. Water may even leak into the cylinder to such an extent that an hydraulic lock occurs when the engine is next started, resulting in a bent connecting rod or other damage.

If, on the other hand, the cylinders and rings are worn, the water may leak into the sump, causing an unaccountable rise in the level of the oil, which will float on top of the water until the engine is run. The oil and water will then form an emulsion, with serious results as far as the bearings are concerned.

Excessive Cylinder Wear

The most common causes of rapid cylinder wear are summarized below—

Incorrect Grade and Viscosity of Oil. Too thin an oil in summer may allow the oil film to break down when the engine is heavily loaded. Too heavy a grade in winter will cause an inadequate oil-flow from the connecting rods for a considerable period after starting from cold ; the cylinder walls are thus starved of lubricant.

Condensation. Condensed petrol will wash the oil film from the cylinder walls. Condensed water, containing acids from

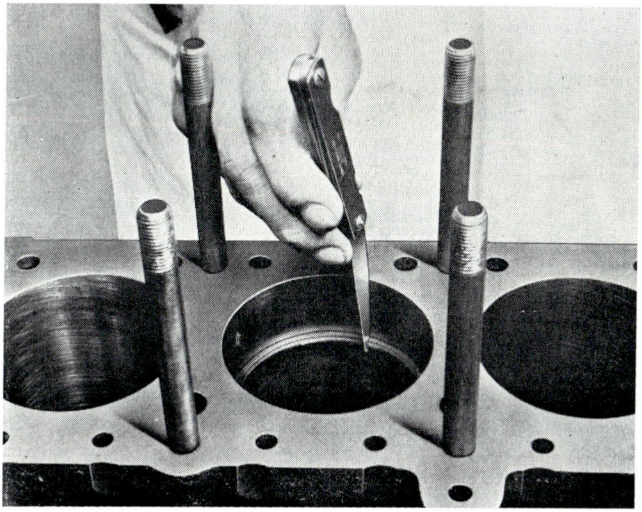

FIG. 17. CHECKING PISTON RING GAP

By measuring the gap at the top of the cylinder (just blow the unworn section) and at the bottom, the degree of cylinder bore wear can be estimated

the products of combustion, will cause corrosion. The combination of the two factors results in rapid wear when the engine is frequently started from cold.

Carburettor Mixture Too Rich. Too rich a mixture or excessive use of the choke will result in surplus fuel condensing in the cylinder walls—see above.

Incorrect Fitting of Connecting Rods. If the connecting rods are provided with oil spray holes, these must face the thrust sides of the cylinders to ensure adequate lubrication.

Incorrectly Fitted Pistons and Rings. Insufficient clearance between the piston and the cylinder will cause partial or complete seizure. If the ring gaps are too small, the rings will butt when hot, causing partial seizure or fracture of the rings; the broken ends of the rings may score the cylinders.

Piston Rings Stuck in Grooves. The leakage of hot gases past the rings will tend to burn away the oil film on the cylinder walls and will also cause oxidation of the metal.

Bent Connecting Rods. Bent rods will impose a side thrust on the cylinder by attempting to twist the piston into line with the rods—see "The Crankshaft and Connecting Rods."

Dirty Air Cleaner. A choked air cleaner may allow abrasive dust to pass into the engine.

Valve Gear Troubles

Enough has been said in preceding sections to admit of effective diagnosis of the various troubles associated with the valves and valve-operating gear, such as gas-leakage, noise, and lack of power and misfiring. Some amplification of the charts will, however, be of value. This involves consideration of factors such as tappet clearance, valve-seatings, mechanical efficiency and valve-timing.

Effect of Incorrect Tappet Clearance. Too great a valve-clearance will cause the valve to open late and close early, while the lift will also be reduced. The result, with an inlet valve is to restrict the flow of mixture into the cylinder, leading to loss of power. On an exhaust valve too great a clearance will cause overheating owing to the restriction on the escape of the exhaust gases, while the valve may be burnt and distorted.

On side-valve engines a wrong impression of the tappet clearance may be given if the foot of the valve-stem has worn a depression in the head of the tappet screw. The feeler gauge will bridge the gap, so that a considerably greater clearance will exist than is indicated by the gauge.

Too small a clearance, on the other hand, will advance the valve-timing. If the clearance disappears when the engine is hot and the valve is held off its seating, spitting back in the carburettor will be experienced. This applies when either the inlet or exhaust valve is at fault. Where the inlet valve is concerned, burning mixture can re-enter the induction manifold ; with the exhaust valve, a proportion of spent gas will be drawn back into the cylinder on the induction stroke, diluting

Fig. 18. This Piston Shows Evidence of Severe " Picking Up " and Scoring of the Skirt, Due to Insufficient Lubrication and Seizure. Heavy Cylinder Bore Wear is Caused by these Conditions

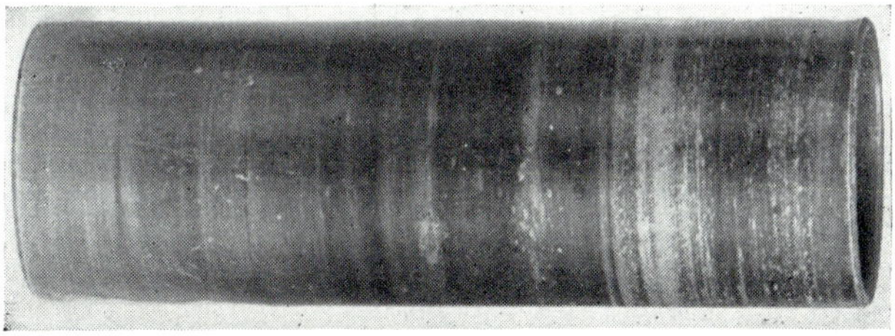

Fig. 19. A Badly-worn and Scored Gudgeon Pin

FIG. 20. CHECKING THE VALVE CLEARANCES ON AN OVERHEAD-VALVE ENGINE

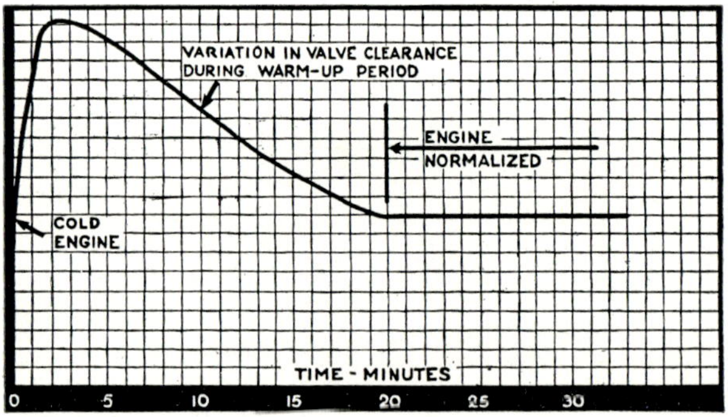

FIG. 21. GRAPH ILLUSTRATING THE VARIATION IN VALVE CLEARANCE THAT OCCURS DURING THE FIRST TWENTY MINUTES AFTER STARTING A PUSH-ROD-OPERATED OVERHEAD-VALVE ENGINE FROM COLD

the incoming mixture and producing a very slow-burning charge, which is still burning when the inlet valve again opens. This point is worth bearing in mind, as lack of exhaust valve tappet clearance or a sticking exhaust valve is frequently wrongly diagnosed as inlet valve trouble, spitting back being normally associated with carburettor and inlet valve faults.

On early Ford engines no tappet adjustment is provided, the correct gap being obtained during decarbonization by grinding

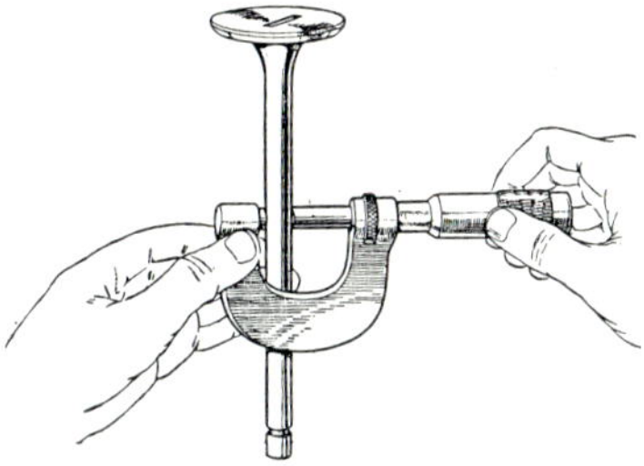

FIG. 22. MEASURING THE DIAMETER OF A VALVE STEM WITH A MICROMETER

the end of the valve stem to increase the clearance, or grinding the valve into its seating to decrease the gap.

Influence of Temperature. The engine temperature considerably influences the valve-clearance. If the clearance is quoted with the engine cold, make sure when checking the clearance, that it really is cold, not slightly warm. Conversely if the clearance must be measured with the engine hot, it should not be measured until the engine has been running for some time, so as to allow all the parts to assume their normal running temperature. This is particularly important on push-rod-operated engines. During the first few minutes of warming-up the valve clearance will increase by several thousandths of an inch owing to the expansion of the cylinder block and head lifting the rockers away from the push-rods, which heat up more slowly. As will be seen from the graph in Fig. 21, twenty minutes at a fast idling speed may elapse before the parts are normalized.

Using a Vacuum Gauge. Mention has already been made of the value of a vacuum gauge as an aid to detecting sticking valves. It is not always appreciated that on overhead-valve engines on which the valve-clearances can be readily set while the engine is idling, the vacuum gauge is a useful aid to accurate setting of the clearances. The clearances should first be set to the nominally correct figure; each adjustment should then be slightly increased or decreased in turn, in an effort to improve the vacuum reading.

If an improvement—that is, a higher vacuum reading—can be obtained only by a fairly substantial variation from the specified figure, it is likely that the particular valve is sticking, and the fault must be rectified.

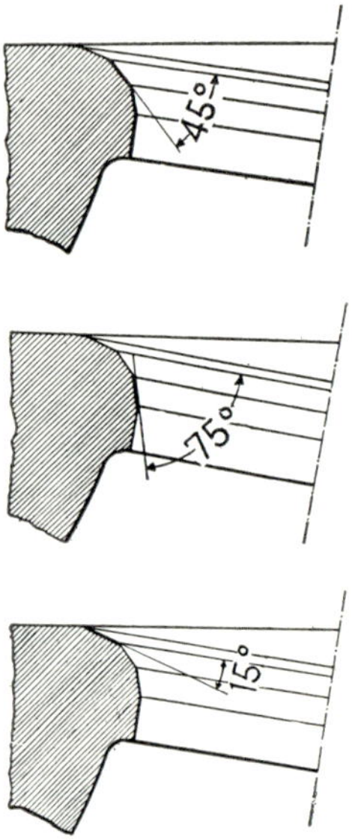

FIG. 23. METHOD OF NARROWING A 45° VALVE SEAT BY USING 75° AND 15° CUTTERS

If a difference of only one or two thousandths of an inch on either side of the nominal figure produces a higher vacuum gauge reading, however, the inference is that slight variations in the camshaft or other parts have been corrected.

Valve and Tappet Troubles. Whenever an engine is decarbonized the valves should be carefully inspected. If the seatings are badly pitted or burnt they should not be ground in; instead, the valve face and the seating in the block should be re-cut with a reseating tool or an abrasive stone. Excessive grinding will result in the valve becoming "pocketed," so that its effective lift is reduced, a cause of power loss which is often overlooked. The use of 75 deg. and 15 or 20 deg. cutters as shown in Fig. 23 will cure this fault.

Before the seatings are re-cut, however, the valve stem must be checked for truth, while both the valve and guide should be tested for wear. Bent stems, and excessive clearance between the stem and guide, will render it impossible to re-cut the seatings accurately.

The tappets and guides should also be checked for wear. On many modern engines the tappets are carried in detachable blocks, so that removal, examination, and renewal of parts is a simple matter. If the tappets are mushroom-shaped and work in bushes in the cylinder block, however, it is generally necessary to remove the camshaft so that they can be withdrawn from below.

Diagnosing Hydraulic Tappet Faults. An alternative to the normal type of tappet which is becoming increasingly popular is the hydraulic tappet, which automatically takes

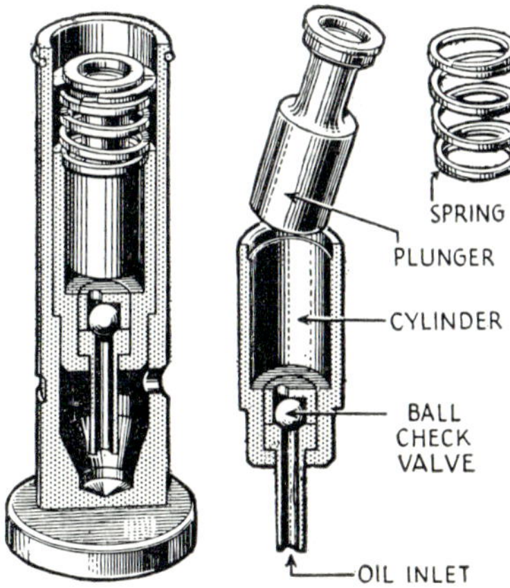

FIG. 24. AN HYDRAULIC TAPPET, SHOWN ASSEMBLED, AND ALSO WITH HYDRAULIC UNIT REMOVED AND DISMANTLED

up all clearance in the valve gear while still allowing the valve to seat properly. These tappets increase or decrease in length during each revolution of the camshaft to compensate for contraction or expansion of the valve mechanism.

As will be seen from Fig. 24, an hydraulic tappet consists of a tappet body, in which is fitted the hydraulic unit. This embodies a plunger which is an accurate fit in a cylinder. The plunger is raised into contact with the valve stem by a spring. Oil from the engine lubrication system enters through a check valve into the cylinder below the plunger, which is thus prevented from sliding back into the cylinder when the tappet is lifted by the cam. A slight leakage of oil past the plunger compensates for expansion of the valve or mechanism, so that the valve will not be held off its seating when the cam leaves the tappet.

Dismantling. Remove the hydraulic unit from the body. Remove the plunger from the cylinder by twisting the plunger and the spring in the direction that would "wind up" the spring, at the same time pulling outwards. If the plunger does not come away easily from the cylinder, it may (in rare cases) be stuck, owing to carbon forming at the shoulder of the cylinder above the plunger.

To determine whether the plunger is stuck, insert a wooden stick in the oil inlet at the bottom of the cylinder to de-seat the ball valve and allow the trapped oil to drain away. If the plunger can now be pressed down, it is not stuck but is held in the cylinder by the carbon deposit. Soak the assembly in clean paraffin and, if necessary, use taped pliers to withdraw the plunger.

The plunger and the cylinder are selectively fitted, and the plunger from one cylinder cannot be used in another cylinder. In servicing a tappet, should it be necessary to discard a plunger or a cylinder, a new pair will be required.

Checking the Leak Down Rate. The rate at which oil escapes between the cylinder wall and the plunger can be checked by making sure that there is no lubricating oil on the cylinder wall or plunger, and holding the cylinder in an upright position. Depress the plunger and release it quickly. The plunger is now operating against air instead of oil, and this air is trapped by the ball-valve and the close fit of the plunger in the cylinder. It should yield slightly to the pressure of the finger on the plunger, but the plunger should kick back upon release of the pressure. If it does not kick back it is due to one of the following conditions—

1. Air is escaping past the ball-valve because dirt is preventing the ball from seating properly.
2. Air is escaping past the ball-valve owing to a damaged ball seat.
3. Air is escaping because the clearance between the cylinder wall and the plunger is excessive.

To determine whether the ball-valve is at fault, place a finger over the oil inlet, press the plunger down and release it quickly. If it kicks back properly the ball-valve is faulty. Wash the unit thoroughly and repeat this test without the finger over the oil inlet. If there is still no kick back, the ball-valve is faulty and the entire assembly (cylinder and plunger) should be discarded.

If there is no kick back when the finger is held over the oil inlet, it is an indication that too much leakage is occurring between the cylinder wall and the plunger ; the entire unit should be discarded.

In the tests mentioned, it is important that pressure be released immediately after depressing. If the plunger is held down too long, all air trapped under the plunger will leak out, defeating the object of the test.

Further Checks. The valve springs should be tested for strength. The simplest plan is to test each against a new spring ; weakened springs will be appreciably shorter, especially when compressed under the same weight. When double valve springs are fitted, the pair should always be renewed together.

Check the rockers and bushes for wear on overhead-valve-engines. It may be possible to renew the bushes alone, but the rocker shaft will also be found to be worn in the majority of cases. The side-play of the rockers should be taken up by fitting suitable shims, or new silencing springs if these are fitted as standard. Any flats on the faces of the rockers should be stoned out.

Summary of Valve Troubles. To sum up the various valve troubles mentioned above, the following diagnosis charts, which also include some items not previously mentioned, may be useful—

Valves Sticking

Insufficient tappet clearance.
Insufficient clearance between guide and stem.
Rocker stuck.
Valve springs weak or broken.
Carbon deposits on stem.
Use of fuel with high gum content.
Hydraulic tappet defective.

Valves Burnt and Pitted

Insufficient tappet clearance.
Excessive carbon deposits.
Valve seats too narrow.
Carburettor mixture too weak.
Weak valve springs.
Valves sticking—see above.
Valve head too thin.
Wrong valve timing.

Checking the Valve Timing

Incorrect valve timing is not a very likely fault unless the timing chain or gears have been dismantled and incorrectly assembled. Sometimes, however, a slack timing chain may jump a tooth on one of the sprockets; this usually occurs when the engine backfires.

Since there is a wide variation in the methods of marking the parts to ensure correct reassembly, detailed instructions are outside the scope of a book of this type. A quick check of the valve-timing without the necessity for removing the timing gear cover, on the other hand, is often useful during fault tracing.

Simple Methods of Checking the Timing. On many engines the appropriate points at which the inlet and exhaust valves should open are marked in the engine flywheel, the letters "I.O" and "E.O" being visible through an inspection aperture. It must be remembered that in some engines the tappets must be set to a special clearance when checking the timing. Moreover the timing must sometimes be checked with the exhaust valve at the peak of its lift, the letters "E.P" then appearing on the flywheel.

For a quick check during a roadside diagnosis, however, advantage can be taken of the fact that, as already mentioned, most modern engines employ some degree of valve overlap. The inlet valve, for instance, may open 5 or 10 degrees before top dead centre, and the exhaust close 5 or 10 degrees after T.D.C. Thus, when the piston is at T.D.C. on the exhaust stroke, both valves will be off their seatings by a similar amount.

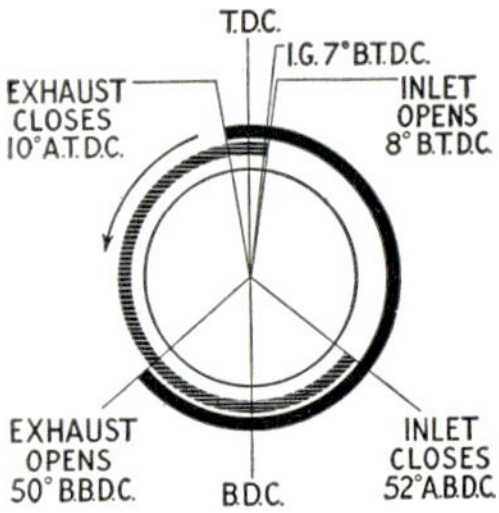

Fig. 25. Typical Valve-Timing Diagram

The piston of the cylinder on which the valves are to be checked—generally No. 1—may be brought to T.D.C. by alining a mark on the flywheel or on the fan pulley with an indicating mark or arrow. Alternative methods of indicating T.D.C. are a plunger which drops into a hole drilled in the flywheel or timing gear, or a rod which is inserted through a hole in the cylinder head, or through a sparking plug hole.

Make sure that the piston in the cylinder on which timing is carried out is on the compression stroke by removing the distributor cover and verifying that the rotor is pointing in the direction of the electrode to which the sparking plug lead serving that cylinder is connected.

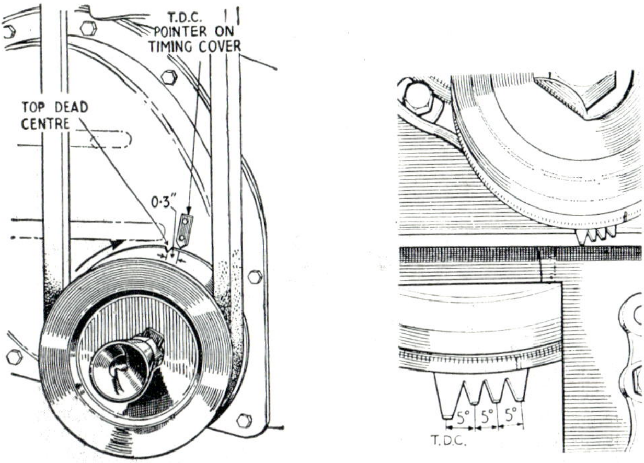

Fig. 26. Timing Mark On Fan Driving Pulley

Sometimes the indicator alines with the lower edge of the pulley and there may be additional pointers giving, say, 5° and 10° advance before t.d.c.

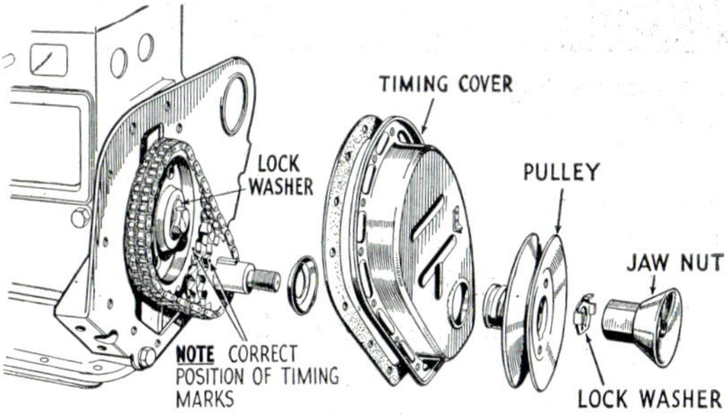

Fig. 27. Timing Chain Sprockets are Also Marked to Facilitate Reassembly

Both valves should now be open slightly or "rocking." While there may be some difference in the degree of opening, it will not be possible for the timing to be incorrect to the

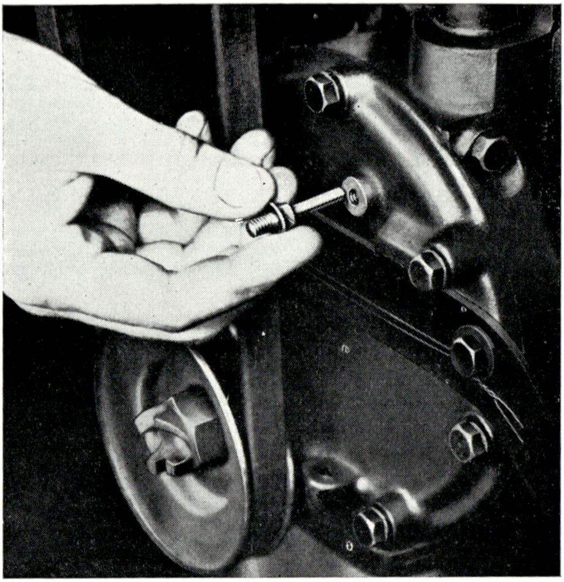

FIG. 28. ANOTHER METHOD FOR FINDING TOP-DEAD CENTRE. ON EARLIER FORD ENGINES, A PIN IS UNSCREWED AND REVERSED, SO THAT IT DROPS INTO A HOLE IN THE TIMING GEAR

extent of a complete sprocket tooth or gear tooth without the error being immediately apparent.

THE CRANKSHAFT AND CONNECTING RODS

Besides diagnosing connecting rod, bearing, and crankshaft troubles as outlined under "Engine Noises" in Chart No. IV further fault-tracing will call for dismantling of the engine. Since the method of carrying out this work depends entirely on the engine, the only advice that can be given here is to obtain the manufacturer's instructions, and particularly a schedule of fits and clearances, without which accurate checking will be impossible. The person armed with this and a micrometer can quickly check excessive wear of the parts.

Apart from wear, however, it is of vital importance to check the alinement of the connecting rods and crankshaft. Often a connecting rod is slightly bent or twisted, and if it is reassembled in this state it will cause rapid cylinder wear owing to the abnormal thrust exerted on the piston. Similarly, a crankshaft may be twisted or "sprung," rendering accurate reassembly impossible.

Since these faults are seldom visible to the eye, the connecting rod should be checked with the aid of a surface plate,

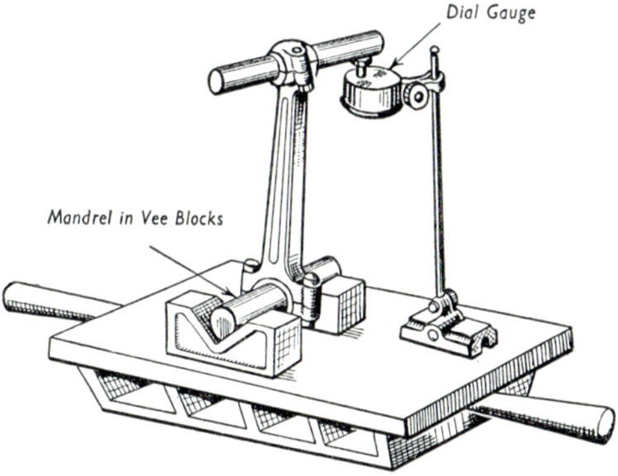

FIG. 29. CHECKING CONNECTING ROD ALINEMENT

vee-blocks, mandrel, and dial gauge indicator, as shown in Fig. 29.

If this equipment is not available, a test can be made with the connecting rod assembled to the crankshaft in the engine. The piston, without rings or gudgeon pin, should be slid down the bore until the gudgeon pin bosses pass on each side of the connecting rod little-end. If the alinement is incorrect, the piston bosses will bind on the connecting rod. This method, of course, will not reveal whether the fault lies in the connecting rod or the crankshaft, but will at least indicate that something is amiss.

The crankshaft can be checked between vee-blocks or between centres in a lathe, again using a dial gauge. If only slightly sprung it can be straightened in a press, provided that heat is not required. The connecting rods can also be set when cold, using a press, bending iron or a vice in conjunction with suitable packing pieces.

Worn or scored main and connecting rod bearings should, of course, be renewed either by remetalling the shells or rods, or by fitting new thin-shell, steel-backed bearings of the type used in many modern engines. The crankpins and main bearing journals must be re-ground if they are badly worn or scored ; undersize bearings will then be required.

The following tables summarize the most usual causes of connecting rod and crankshaft troubles—

Connecting Rod Bearing Failure

Low oil pressure.
Lack of oil.
Improper grade of oil.
Bent connecting rod.
New bearings fitted to rough or oval journals.
Bearings too tightly fitted.
Bearing shells loose in connecting rod.
Clearance too great—bearings hammer out.
Bearings not correctly run-in.

Crankshaft Bearing Failure

Low oil pressure.
Lack of oil.
Incorrect grade of oil.
Crankshaft out of alinement.
Bearings not bored or scraped in line.
Oval or scored journals.
Loose bearing shells.
Bearings fitted too tightly.
Bearings not correctly run-in.

Lubrication Troubles

Used in conjunction with the following notes, Chart No. 5 should enable lubrication troubles to be diagnosed with the minimum of difficulty. Usually, lubrication troubles are caused by, or associated with, other engine faults, so that cross-reference to preceding charts will be necessary in some instances.

Excessive Oil Consumption. Apart from wear of cylinders and piston rings, a cause of heavy oil consumption which is often overlooked is excessive clearance of the piston rings in their grooves. Although the rings may form a gas-tight seal against the cylinder walls, their vertical movement in their grooves causes them to act as small oil pumps. Fig. 31 shows how the oil which is scraped from the cylinder walls during the downward stroke is transferred to the top of the ring and injected into the combustion chamber on the upward stroke.

Worn valve guides on overhead-valve engines will also allow

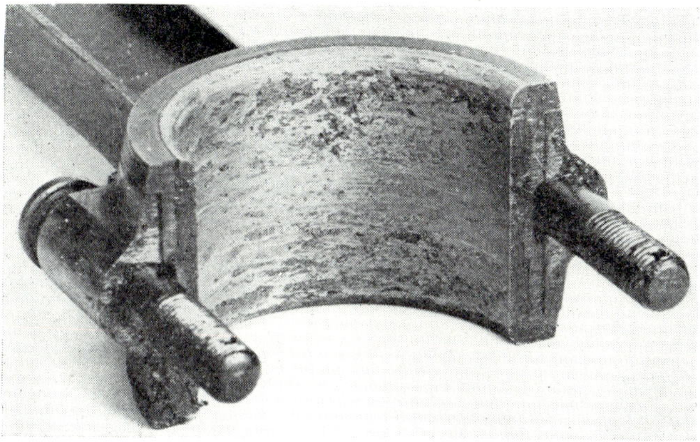

FIG. 30. CONNECTING ROD BEARING THAT HAS FAILED OWING TO LACK OF LUBRICATION

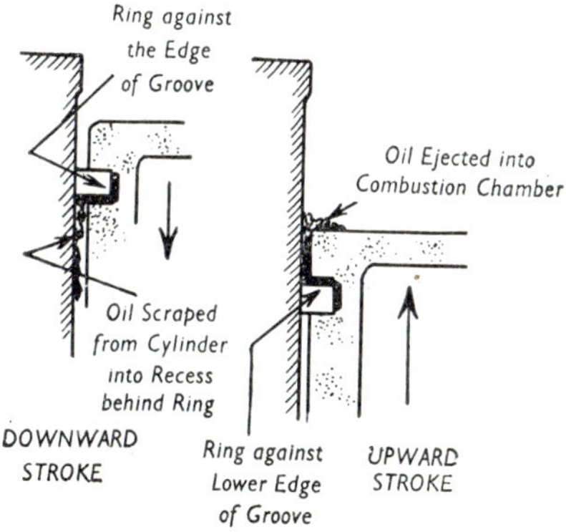

FIG. 31. THE PUMPING EFFECT OF SLACK PISTON RINGS

CHART No. 5

ENGINE LUBRICATION TROUBLES

EXCESSIVE OIL CONSUMPTION
- Check for cylinder and piston ring wear (see "Poor Compression" and "Engine Noises," Chart No. 4).
  - Symptoms of piston and ring wear or blow-by
    - Worn cylinders and rings.
    - Scored cylinders.
    - Broken rings.
    - Ring gaps in line.
  - Compression normal
    - Excessive vertical ring clearance in grooves.
    - Worn valve guides.
    - Defective oil seals on valve stems (o.h.v. engines)
    - Leakage past cylinder head studs into inlet manifold.
    - Badly worn connecting rod bearings.
    - External leakage.

LOW OIL PRESSURE
- Verify oil level in sump and that correct grade of oil is in use.
  - Check pressure with independent gauge
    - If correct
      - Faulty oil pressure gauge or indicator.
      - Faulty or blocked pipe from engine to gauge.
    - If incorrect
      - Clean and adjust relief valve
        - If pressure improves
          - Grit on valve seating.
          - Weak or incorrectly adjusted spring.
        - No improvement in pressure
          - Valve ball or plunger ridged.
          - Worn main and/or connecting rod bearings.
          - Suction filter clogged.
          - Excessive flow to valve gear.
          - Worn oil pump.
          - Broken or leaking oil pipe.
          - Air leaks at flanges, unions or filters on suction side of pump.

OIL PRESSURE TOO HIGH
- Verify grade of oil: too heavy a grade will cause excessive pressure in winter
  - If correct grade in use, check pressure with independent gauge (see above).
    - If incorrect, check relief valve adjustment
      - Incorrectly adjusted relief valve.
      - Choked oil feed—probable cause of bearing failure.
      - Choked oil pipe.
      - Restricted supply to o.h.v. gear

oil to enter the combustion chamber. This trouble is often mistaken for cylinder and piston ring wear, as the amount of oil that enters may be appreciable, producing a very smoky exhaust and causing the plugs to oil up.

To diagnose the trouble, allow the engine to idle for a few minutes, and then open the throttle wide. If clouds of blue

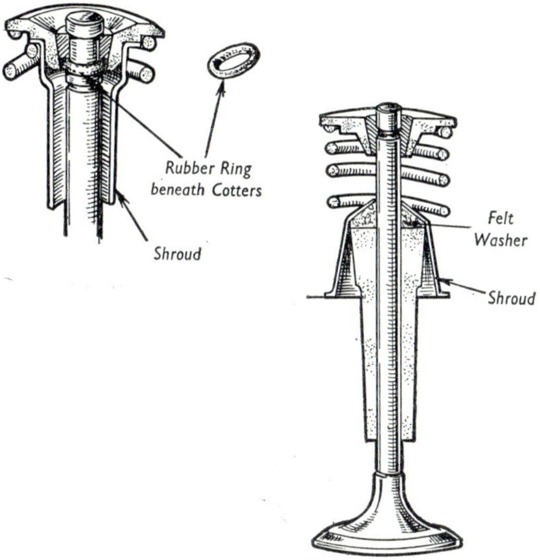

FIG. 32. METHODS OF PREVENTING OIL LEAKAGE PAST OVERHEAD-VALVE STEMS

smoke are ejected from the exhaust, but the smoke quickly diminishes or disappears, the valve guides can be suspected.

If, on the other hand, the trouble is due to worn cylinders and rings, the blue smoke generally increases at high revs owing to the increased oil flung from the connecting rod bearings. Incidentally, if the bearings are badly worn they may allow so much oil to be thrown into the cylinder bores at high engine speeds that the rings may be unable to control it, although the latter may be in a reasonable condition.

Shrouded Valve Guides. To revert to the question of leakage past the valve stems : a number of modern overhead-valve engines have tubular shrouds fitted to the end of the valve stem or to the upper end of the valve guide. In both cases the effect is to prevent excess oil reaching the valve stem. If the oil seal (on the first type of shroud in the form of a synthetic rubber ring, or on the second design a felt washer), is defective, the efficiency of the device will naturally be reduced.

Metering the Oil Supply to Overhead Valves. The flow of oil to an overhead rocker shaft or camshaft is generally metered either by a restrictor pin inserted in the oilway or by an

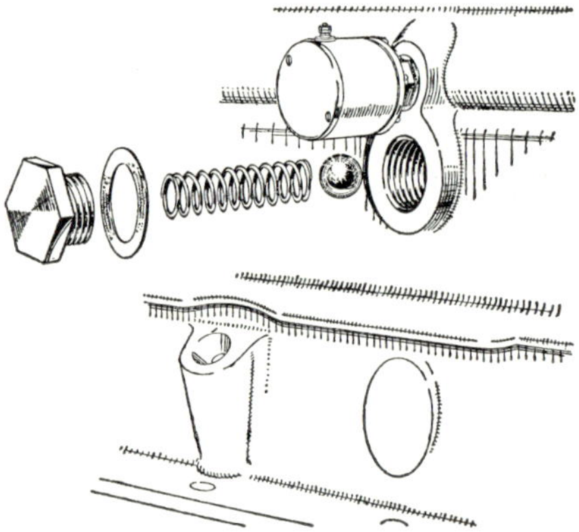

FIG. 33. A NON-ADJUSTABLE TYPE OF OIL PRESSURE RELIEF VALVE

adjustable metering screw. If the oil exudes from the rockers or camshaft bearings in greater quantities than a slow drip at idling speeds, the restrictor should be examined and if necessary adjusted. Sometimes fabric wicks are inserted in the oilways drilled in the rockers or camshaft, and these may need renewal if the oil supply is excessive.

If too much oil is allowed to flow over the top of the cylinder head, some may find its way down studs which pass through the inlet ports. In most cases the nuts fitted to such studs have copper washers beneath them, but these may have been overlooked or replaced by steel washers during assembly.

External Leakage. Oil may leak from the engine at a number of points, but such leakage is fortunately obvious in most cases, and can usually be remedied by renewal of the appropriate gaskets or oil seals, or by tightening a slack union.

When leakage is experienced from the front or rear main crankshaft bearing housings, the trouble may be due to excessive bearing wear or, more simply, to a defective oil seal at this point. Very often a packing ring is inserted in a groove in the sump which surrounds the crankshaft. It is advisable to replace this packing whenever the sump is taken off. The new packing may render the crankshaft stiff to turn at first, but if the front sump bolts are left slightly slack and then tightened progressively while the engine is running, the packing will quickly bed-in.

Low Oil Pressure. If the sequence shown in the chart is followed it should not be difficult to trace the cause of low oil pressure. Sometimes the oil pressure relief valve gives trouble. When it is accessible from outside the engine (fitted, for instance, in a boss on the crankcase), it is a simple matter to unscrew the screw or plug which retains the spring and the ball or plunger, so that these parts can be withdrawn for inspection and cleaning.

If the ball or the seating of the plunger is pitted or ridged, renew the parts. The seating in the crankcase may be cleaned with a non-fluffy rag moistened with petrol or paraffin, after the engine has been rotated a few times by means of the starter to induce a flow of oil from the valve orifice which will wash away any particles of carbon or grit. When the valve is reassembled, the correct oil pressure can be obtained by means of the adjusting screw. On non-adjustable valves, if the spring has weakened, the spring tension can be increased by fitting a washer between the spring and the interior face of the plug. To lower the pressure on this type of valve, fit one or more additional washers between the external flange of the plug and the crankcase.

When the relief valve is incorporated in the oil pump it is generally necessary to remove the sump, and possibly a filter surrounding the pump, to obtain access to the valve ; otherwise the procedure just described applies. When the spring-retaining screw is locked by a nut, however, do not screw the nut along the thread of the screw, or the original setting will be lost. This will entail again removing the sump to readjust

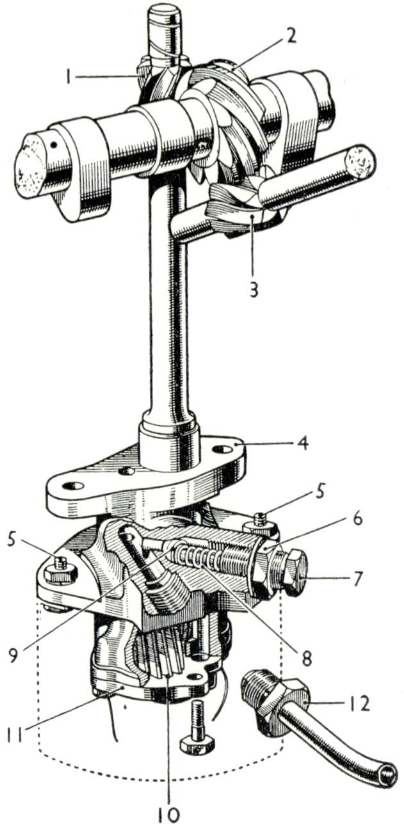

FIG. 34. LIKELY OIL PUMP TROUBLES. ON SOME ENGINES THE DISTRIBUTOR DRIVING SHAFT IS DRIVEN DIRECTLY FROM THE OIL PUMP SHAFT, INSTEAD OF BY A SEPARATE SHAFT AS SHOWN

1. Worn oil pump driving gear
2. Worn skew gear on camshaft
3. Distributor driving gear worn or incorrectly meshed, giving wrong ignition timing
4. Oil pump body cracked or fractured
5. Filter retaining screws loose
6. Pressure relief valve lock nut and washer loose
7. Relief valve adjustment incorrectly set
8. Weak spring
9. Dirt on valve seatings ; ridged seating faces
10. Worn pump gears
11. End cover worn, allowing excessive end-float on gears
12. Loose delivery pipe union ; choked pipe

the spring tension if on reassembly the oil pressure is found to be incorrect.

High Oil Pressure. An unduly high oil pressure should be regarded with suspicion rather than with satisfaction. A pressure higher than usual will be recorded, of course, for some time after starting from cold, but the pressure should drop back to normal as the engine warms up. Provided that the pressure relief valve is correctly adjusted, too high a pressure can be due only to the oil flow being restricted at some point, the inference being that some vital part is being starved of lubricant.

Likely points to check are the restrictor or adjuster in the oil feed to the overhead-valve gear mentioned earlier ; the former may be choked by an accumulation of grit on the flange, while the adjustable type of restrictor often has a tendency to work towards the closed position, gradually shutting off the supply.

If these points are in order, it is advisable to remove the sump so that the oil feed pipes or drillings to the crankshaft may be checked. This may be a lengthy business, entailing a good deal of dismantling. Before going to this trouble, therefore, it is as well to drain the sump, put in a special flushing oil, and run the engine in an attempt to dislodge or flush away the obstructing material.

If a bearing has previously failed and been renewed, the most likely cause of high pressure is the entry of particles of white metal into the crankshaft oilways. This fault will account for the repetition of bearing failures which is sometimes experienced.

CHAPTER IV

TRACING IGNITION TROUBLES

ALTHOUGH the simple ignition tests outlined at the beginning of Chart No. 1 are sufficient to enable one to differentiate between ignition and carburation faults, a more exhaustive series of tests will be necessary to trace the defective component when difficult starting or misfiring can be attributed to the ignition system.

The various types of proprietary testing equipment on the market enable any part of the ignition system to be checked quickly and accurately. Depending on their complication and cost, they embody all or some of the following instruments : voltmeter, ammeter, ohmmeter, master ignition coil, master condenser, motor-driven contact-breaker, condenser tester, neon lamp, and "sparkmeter."

Since the ignition system must often be tested and overhauled without the aid of such comprehensive equipment, however, it is proposed in this chapter to describe simpler fault-tracing methods which can be carried out with the minimum of equipment. In any case, the manufacturers of a test set of the type mentioned generally provide a comprehensive handbook which details the methods of carrying out the various tests with the particular instruments incorporated in the set.

Chart No. 6 will serve as a guide to the sequence of tests, so that we can confine ourselves to the actual tests and checks.

Checking the Low-tension Circuit. Let us suppose that when the high-tension lead is removed from the distributor cap and held about $\frac{3}{8}$ in. away from an earthed metal surface, while the contact breaker points are flicked apart with the ignition switched on, no spark is obtained. As will be seen from the chart, the cause may lie in the low-tension circuit, the high tension lead itself, or the coil. It is a simple matter to substitute a short length of wire for the high-tension lead and repeat the test, so that this possibility can be quickly confirmed or ruled out.

The most likely alternative is a fault at some point in the

low-tension, or primary, circuit. As will be seen from the wiring diagram, this can be traced from the fuse box to the ignition switch, and thence through the coil to the contact-breaker, where the current passes through the contact points to " earth."

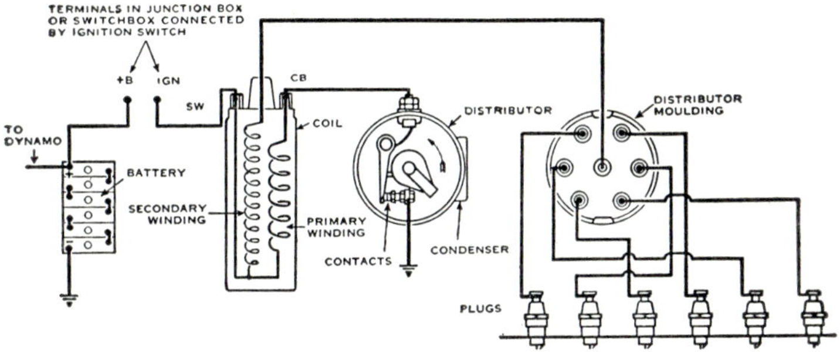

FIG. 35. THE CIRCUIT USED IN A COIL IGNITION SYSTEM FITTED TO A SIX-CYLINDER ENGINE

If the self-starter works satisfactorily and the lights are bright, the battery can be taken to be in order ; otherwise, the wiring and connexions between the battery and fuse and junction box must be checked as described in Chapter XIV.

For the present, however, the focus is on the primary circuit of the ignition system. If an ammeter is fitted to the car, the continuity of the primary circuit can be verified by the fact that the needle registers 2–3 amps on 12 volt circuits, or 3–4 amps on 6 volt systems, whenever the contact breaker points are closed. If an ammeter is not fitted, a simple tester should be made up from an S.B.C. lampholder fitted with a 12 volt, 24 watt, or a 6 volt, 18 watt bulb and a pair of flexible leads, to the ends of which are attached small crocodile clips.

Attach one clip to the low-tension terminal on the side of the contact-breaker and the other to a bright earthed surface on engine or chassis. The lamp should light-up whenever the contact points are separated and should go out when they are closed. If the lamp fails to light, or glows dimly, a disconnexion or high resistance at some point in the low-tension circuit is indicated. Work back towards the battery, cleaning

CHART No. 6

TRACING IGNITION TROUBLES

This chart deals only with ignition system faults. Other causes of misfiring, loss of power, and difficult starting, are covered in Charts Nos. 1, 2, 3 and 4.

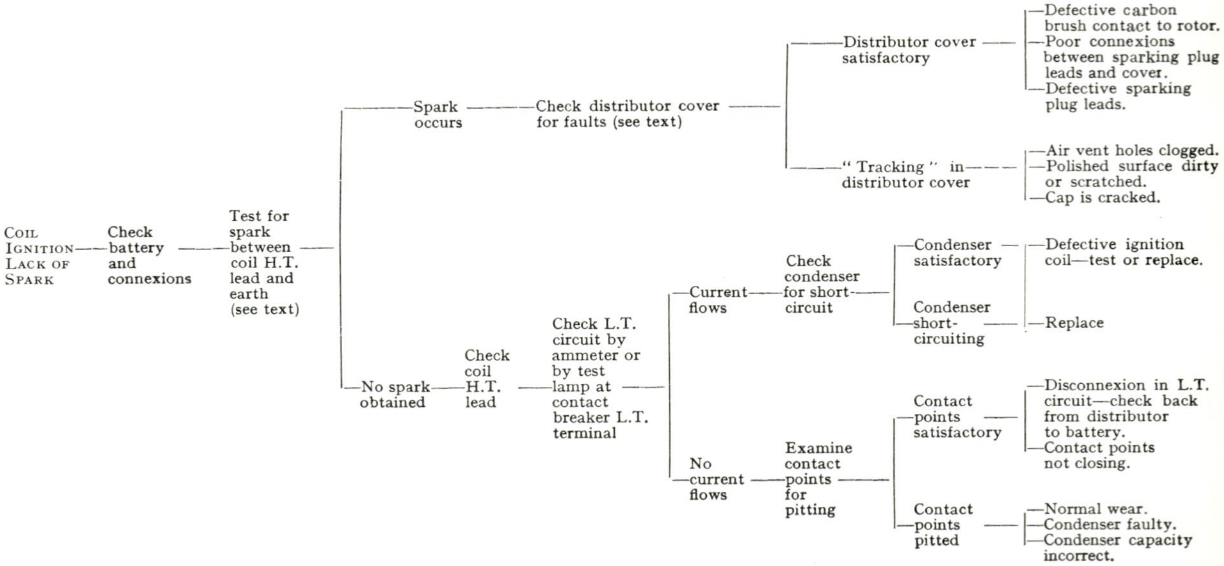

Coil Ignition Misfiring

- Check battery and connexions
  - Check condition of contact points and gap
    - Points satisfactory and gap correct
      - Check L.T. circuit for poor connexions or intermittent disconnexion or short-circuit
        - If satisfactory
          - Poor earth connexion between distributor and engine or engine and chassis—fit earthing lead.
          - Coil faults.
          - Distributor faults.
          - Wrong ignition timing.
          - Automatic advance mechanism inoperative.
          - Sparking plug faults (see Chart No. 7).
    - Points pitted
      - Test condenser for open circuit or wrong capacity.
    - Gap does not remain same when tested on each lobe of cam
      - Cam spindle bearing worn.
      - Contact breakers not synchronized on twin-rocker arm distributors.

Magneto—Lack of Spark and Misfiring

- Test for spark at slip ring (see text)
  - Satisfactory spark occurs
    - Examine distributor brush
      - If satisfactory
        - Distributor faulty.
        - Rotor brush broken stuck.
        - Electrodes burnt.
        - Sparking plug leads defective.
        - Sparking plug troubles (see Chart No. 7).
        - Timing incorrect.
  - No spark obtained or weak spark
    - Examine contact breaker points and check gap
      - Points satisfactory and gap correct
        - Test armature (see text)
          - Armature satisfactory
            - Magnet weak.
            - Short-circuited condenser.
          - Armature faulty
            - Broken or chafed windings.
            - Armature rubbing on pole pieces.
      - Points badly pitted
        - Condenser faults

and tightening each terminal, and making a test at each stage : the " C.B." terminal on the coil ; then the " S.W." terminal ; next the coil side of the ignition switch, followed by the " live " side, and so on.

It is as well to move the wires about when making the test to show up internal fractures which may be evident only when the wire is moved. If the lamp does not glow when the " C.B." terminal on the coil is touched, for instance, but does so when connexion is made to the " S.W." terminal, the

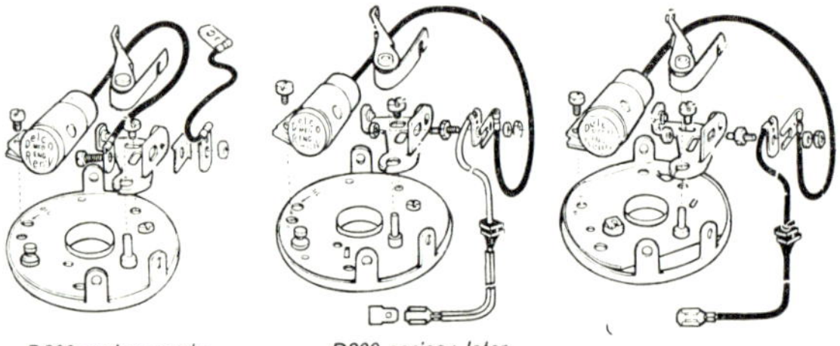

FIG. 36. THREE DELCO-REMY DISTRIBUTOR CONTACT-BREAKER ASSEMBLIES

inference is that there is an open circuit in the primary winding of the coil.

Faulty Contact Points. To revert to our test with the lamp connected between the contact breaker low-tension terminal and earth. If the lamp does not go out, or if it glows with reduced brilliance when the contact breaker points are closed, the trouble is due to dirty or pitted contact points. It is a simple matter to true up the points with a fine file, providing that the surfaces are kept dead square, but by far the best procedure is to renew them. Modern tungsten points are not expensive, and should preferably be renewed whenever a new set of sparking plugs is fitted—say every 10,000 miles.

Testing the Condenser. If the points become badly pitted after only a short period, a faulty condenser is a likely culprit. This component can also be suspected if the test lamp does not light-up, or the ammeter continues to show a reading, when the contact points are open, the inference being that current is escaping to earth through a short-circuit in the condenser.

Disconnecting the condenser will immediately reveal the soundness or otherwise of the diagnosis.

It is difficult to test a condenser for faults other than a short circuit without the use of specialized equipment. Most proprietary test sets incorporate a source of direct current, either in the form of a small generator or a rectifier for use on A.C. mains. By this means the condenser can be charged, and its capacity to hold the charge judged.

If *direct current* is available from the mains, preferably at a voltage of at least 200 volts, an ordinary neon lamp as sold for household use can be fitted in a holder connected to one side of the supply. To the second contact of the lampholder attach a test lead. A second test lead is connected to the other pole of the mains supply. The two leads are then placed in contact with the terminal and the case of the condenser and the current is switched on.

The neon lamp should flash once as the condenser is charged. If the condenser holds its charge perfectly the lamp will not flash again. In practice, some leakage generally occurs, but if the lamp does not flash more frequently than two or three times a minute the condenser can be considered to be in good condition. If D.C. mains are not available, a pair of 120 volt wireless batteries can be connected in series to provide the current. They will have a long life, since the current consumption during the test is very small.

Instead of test equipment being employed, a condenser which is known to be in perfect condition may be substituted for the one which is suspect. If this eliminates the misfiring or other trouble previously experienced, it can be assumed that the original condenser is faulty. A replacement condenser is inexpensive, so that it is questionable whether the outlay on test equipment of the type described is justified unless a large number of condensers are to be tested.

It should be emphasized that it is essential to fit a replacement condenser of the correct capacity : this may be between 0·1 and 0·4 mfd, the value depending on the characteristics of the ignition system. To determine whether a condenser of the correct type is fitted, examine the contact breaker points after several thousand miles of running. If the positive pole of the battery is earthed, as was usual until recently on British cars, the formation of a pit or crater in the

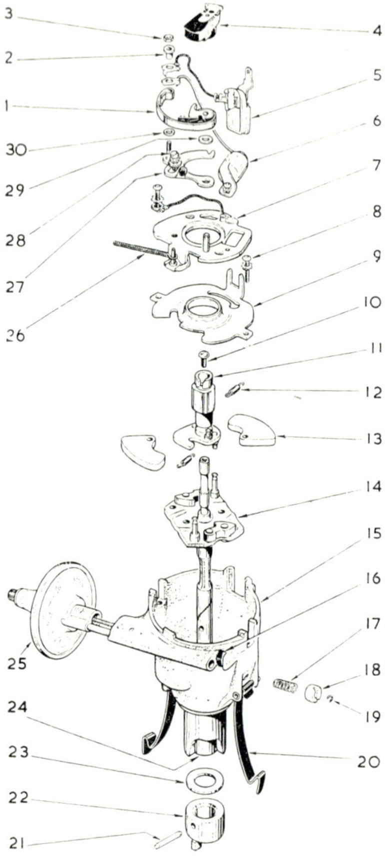

FIG. 37. TYPICAL LUCAS IGNITION DISTRIBUTOR

1. Spring contact and rocker arm
2. Insulating sleeve
3. Nut
4. Rotor arm
5. L.T. terminal
6. Capacitor (condenser)
7. Contact plate
8. Screw
9. Base plate
10. Screw
11. Cam
12. Centrifugal spring
13. Centrifugal weights
14. Action plate and shaft assembly
15. Distributor body
16. Ratchet spring
17. Coiled spring
18. Adjusting nut
19. Circlip
20. Cap retainer
21. Pin
22. Driving dog
23. Washer
24. Bearing sleeve
25. Vacuum-advance unit
26. Vacuum connecting spring
27. Fixed contact
28. Fixed contact plate locking screw

29, 30. Insulating washers

surface of the point attached to the stationary plate, and a corresponding "pip" or projection on the point carried by the pivoted arm, suggest that the condenser has too high a capacity. Conversely, a crater in the point attached to the

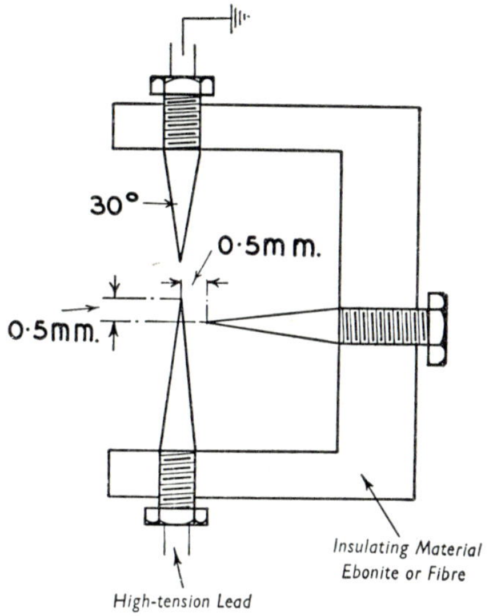

FIG. 38. STANDARD THREE-POINT TEST GAP

rocking-arm indicates under-capacity. If a negative-earthed system is used, the indications will naturally be reversed.

Testing the Coil. As with the condenser, the quickest way to check the coil is to substitute a replacement of known efficiency. The test for a spark from the high tension lead to earth, as already described, gives a fair indication of the efficiency of the coil. If it is necessary to decrease the spark gap to, say, $\frac{1}{16}$ in. before a spark occurs, the coil is obviously weak, assuming that the other components in the low-tension circuit have been checked.

Most non-electronic test kits include a standard three-point spark gap for testing coils. This is shown in Fig. 38. The brass electrodes should have nickel tips, and the points must be kept sharp, as their contour affects the voltage necessary to cause a spark to jump the gap. The adjustment of the third, static electrode is also important, as it controls the degree of

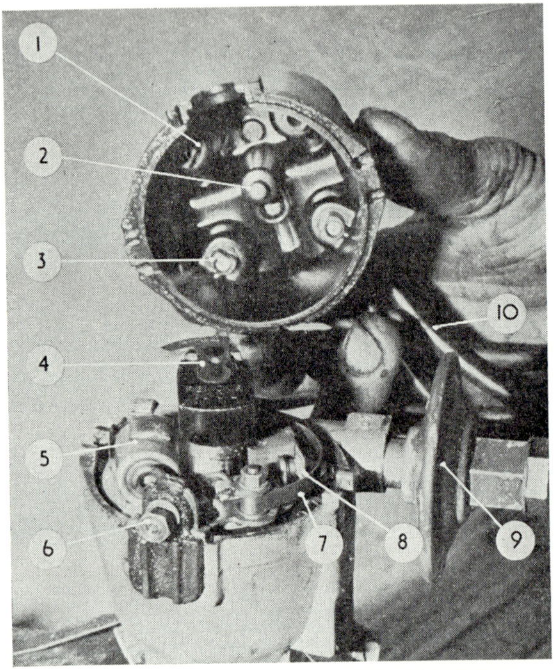

FIG. 39. CONTACT-BREAKER AND DISTRIBUTOR FAULTS

1. "Tracking" in distributor cap
2. Broken or sticking central brush
3. Corroded electrodes
4. Burnt electrode or cracked rotor
5. Defective or badly earthed condenser
6. Loose or dirty terminals
7. Weak or broken spring
8. Burnt, dirty or incorrectly adjusted contacts
9. Faulty timing control
10. Perished leads

ionization across the main gap, and thus the voltage of the spark.

On normal ignition equipment the main gap should be set to $5\frac{1}{2}$ mm. For high-voltage and "sports" coils, the gap should be $7\frac{1}{2}$ mm. A voltage of 11,000 volts is required to produce a spark across the $5\frac{1}{2}$ mm gap ; the $7\frac{1}{2}$ mm gap calls for 14,000 volts.

It is important to check the performance of the coil by passing current through it for some time after it has been allowed

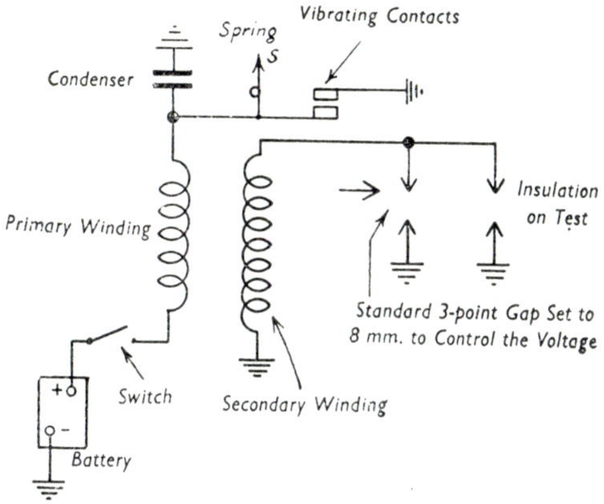

FIG. 40. A TEST CIRCUIT FOR CHECKING THE INSULATION OF THE DISTRIBUTOR CAP AND OTHER PARTS

to warm up, as an elusive fault can be caused by an internal fracture which causes an open circuit or short-circuit only when the coil is hot.

Distributor Cover Faults. The continuous sparking which takes place between the distributor rotor and the electrodes in the cap causes the formation of ozone, owing to the ionization of the air across the gap. This leads to the deposition of nitrous powder on the interior of the cap, and of nitric acid on the electrodes, if condensed moisture is present. Apart from corrosion of the electrodes, the nitrous powder is a conductor of electricity, and will lead to " tracking " of the spark between the electrodes. This leakage of current is revealed by a zig-zag black line, which may be mistaken for a crack.

The cap should be cleaned with water, which dissolves the nitrous deposit, and the electrodes should be scraped clean. If an induction coil is available, as fitted in many test sets, the cap can then be tested. An ordinary ignition coil energized by a motor-driven contact breaker on the test bench will serve.

A standard spark gap of the type just described should be connected between the high-tension terminal of the coil and earth to control the voltage, and should be set to an 8 mm gap to provide a voltage of 15,000 volts at the test prods. These should be touched to pairs of electrodes in turn, when any tracking will be revealed by a continuous stream of sparks.

Tracking is a cause of misfiring on one or two cylinders, which is often overlooked. The resistance at the plug points in a cylinder which is on the compression stroke is much higher than that across the leakage path in the cap. As a result the spark takes the least line of resistance and jumps across to another electrode. The only permanent cure is to fit a new cap; as a temporary measure scrape the track carefully until clean plastic is exposed and polish with fine glasspaper.

Locating Magneto Troubles

It is a simple matter to apply a preliminary test to a magneto without dismantling it; the low-tension and high-tension circuits, with the exception of the distributor and sparking plug leads, are included.

On the usual rotating-armature type of magneto the high-tension current is collected by a carbon brush which makes contact with a slip ring at the rear of the magneto. This is revealed on removing a cover in the vicinity of the driving sprocket, and it is here that the first test is made. At this stage faults such as broken or stuck collector and distributor brushes and a dirty, burnt, or pitted slip ring will be immediately obvious.

Remove the collector brush-holder and rest the blade of a screwdriver against the metal body of the machine so that the tip of the blade is $\frac{1}{4}$ in. away from the metal of the ring. Alternatively, attach a length of stiff wire to an earthed bolt or screw and bend it to give the correct gap. When the magneto is rotated, regular sparks should jump the gap.

If no spark is obtained, the fault may lie in the low-tension circuit, including the contact breaker and condenser, in the high-tension wiring, or in a weak magnet. Before investigation proceeds, however, the contact points should be examined and cleaned or filed up if necessary, and the earthing bush fitted on many magnetos should be checked. Also make sure that the rocker arm is not binding on its pivot.

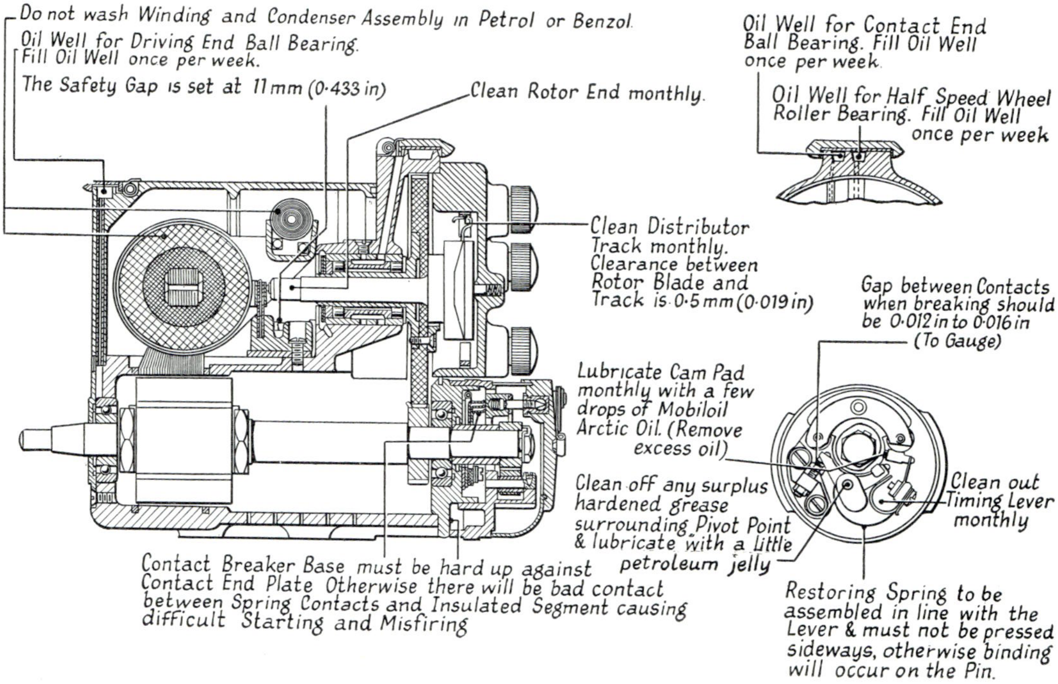

FIG. 41. TYPICAL SERVICE CHART FOR AN INDUCTOR-TYPE MAGNETO

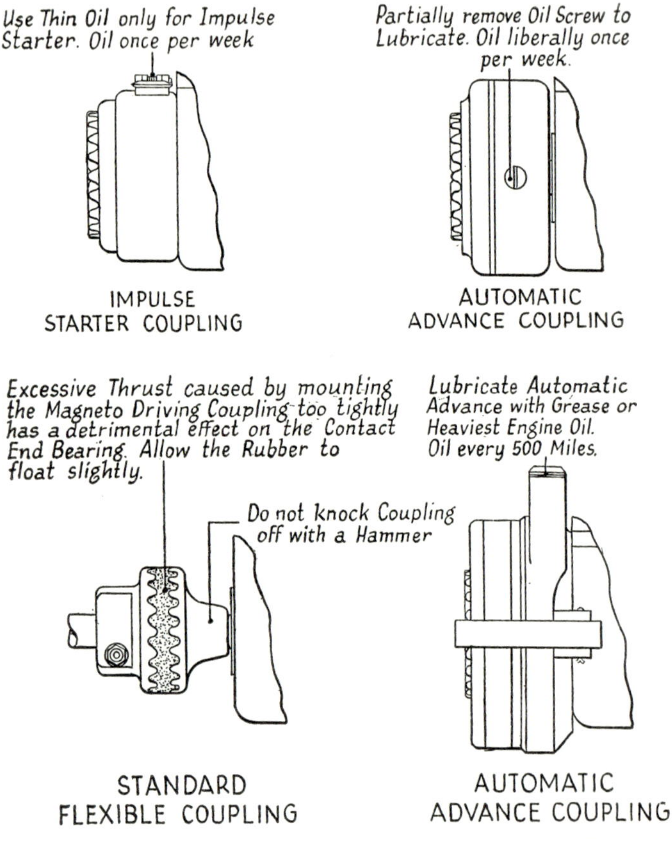

FIG. 41 (*contd.*). TYPICAL SERVICE CHART FOR AN INDUCTOR-TYPE MAGNETO

Armature Tests. Normally, armature tests are best carried out with the armature removed, but a preliminary test can be applied with it in place. Arrange a length of stiff wire under

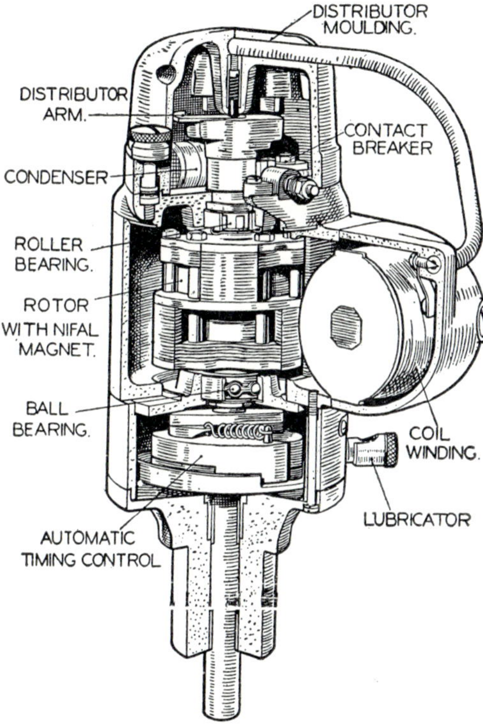

Fig. 42. Lucas Vertical-type Magneto, Designed to Replace a Coil-ignition Distributor

an earthed nut or screw to give a $\frac{1}{4}$ in. gap at the slip ring as before. One terminal of a six-volt battery should be connected by a wire to the frame of the magneto, and a wire attached to the other terminal should be touched against the centre screw of the contact breaker, making sure that the points are open, and holding the driving spindle so that the armature cannot rotate. If no spark is obtained across the test gap, an armature fault or a short-circuited condenser may be suspected.

To discover where the trouble lies, the armature must be removed from the magneto for a thorough test on an electrical test bench, including checking between centres, and, if necessary, rewiring. This work, however, is best left to a magneto specialist, though the magnet can be remagnetized without dismantling the magneto when the necessary equipment is available.

Magneto Distributor Faults. If a spark can be obtained at the slip ring, it follows that the fault must lie in the distributor or sparking plug leads. Points to check are the distributor brush in the collector brush housing, the brush or rotor which serves the distributor segments, and, of course, the sparking plug leads. On wipe-brush distributors make sure that the insulation between the metal segments is not scored and that a greasy carbon track does not short-circuit the segments. If "tracking" of the current takes place, the surface of the insulator will quickly be burnt and a replacement distributor cap will usually be the only effective cure.

Inductor-type Magnetos. On many modern magnetos, especially on the vertical type designed to replace the coil ignition distributor on sports engines, a rotating permanent magnet replaces the armature, and the windings and condenser are stationary. On vertical magnetos the contact-breaker and distributor resemble the components used on coil ignition systems, so that, as far as these parts are concerned, fault-tracing is more on the lines described earlier in this chapter.

Checking the Ignition Timing

It is a simple matter to check the timing of a coil ignition system when an ammeter is fitted to the car, since the ammeter needle will drop back to zero when the contact points open. It is necessary, therefore, only to bring the timing mark in the flywheel slowly into register with the indicator, while an assistant watches the ammeter. Unfortunately this may involve three people : one to turn the engine by the starting handle or by pushing the car with top gear engaged, one to observe the timing mark, while the third watches the ammeter.

The best plan is to make use of a timing light in the form of a sidelamp bulb in a holder fitted with two leads, connected, as described earlier in this chapter, to the low-tension terminal on the contact-breaker and to an earthed point on the engine or chassis.

The engine should be set to top dead centre by the particular method provided, as described under "Checking the Valve Timing" in Chapter III, and the pinch bolt which secures the clamp beneath the distributor should be slackened. If the sparking plugs should not fire at T.D.C., the flywheel must, of course, be brought to the correct point before or after T.D.C. Always bring it to the required point by rotating it forwards to take up the play in the timing gears and distributor drive.

Switch on the ignition and rotate the distributor slowly in either direction. As the points break, the bulb will light up; when they are closed, they short-circuit the bulb. Thus a critical setting can be obtained. The distributor clamping bolt should then be re-tightened.

A similar method can be employed to time a magneto. Current must be supplied by a dry battery, a torch bulb of suitable voltage being used. On rotating armature magnetos, the central retaining screw should be removed from the contact-breaker, which will remain in position owing to the wedging action of its taper. The test leads may then be connected to each contact. The bulb will go out in this case when the points separate—the opposite state of affairs to the coil ignition distributor.

The Effect of Contact-Breaker Gap. Besides influencing the intensity of the spark, the contact-breaker gap will also affect the ignition timing. Increasing the gap will advance the timing; decreasing it will retard the spark.

When the distributor is fitted with twin rocker-arms it is essential that the gaps should be equal in order to synchronize the two sets of points. There are two types of distributor employing this arrangement. The first, now becoming less common, uses two rocker arms in parallel, the object being to provide double the contact area and so reduce the resistance in the primary circuit.

The second type of distributor is fitted to multi-cylinder engines; here the two sets of points open and close alternately,

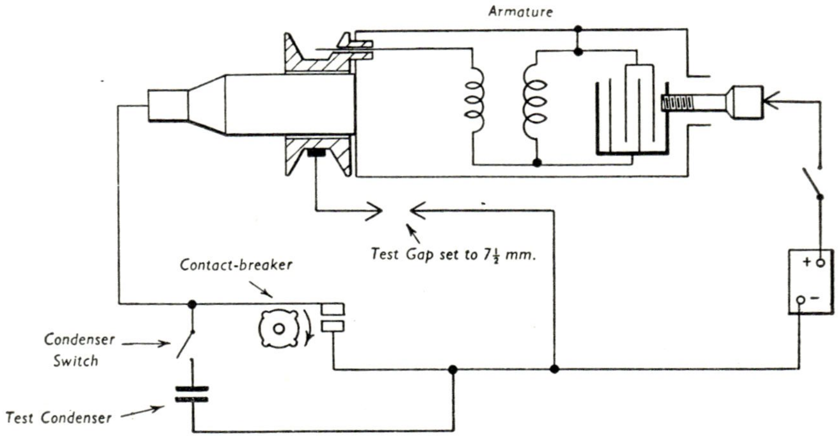

FIG. 43. MAGNETO ARMATURE TEST CIRCUIT

each being responsible for firing half the total number of cylinders. This can be recognized by the fact that the number of lobes on the cam is equal to half the number of cylinders. If an engine fitted with this type of distributor misfires on three, four, or six cylinders, therefore, as the case may be, the trouble can be attributed to incorrect synchronization of the rocker arms. Usually the work can be carried

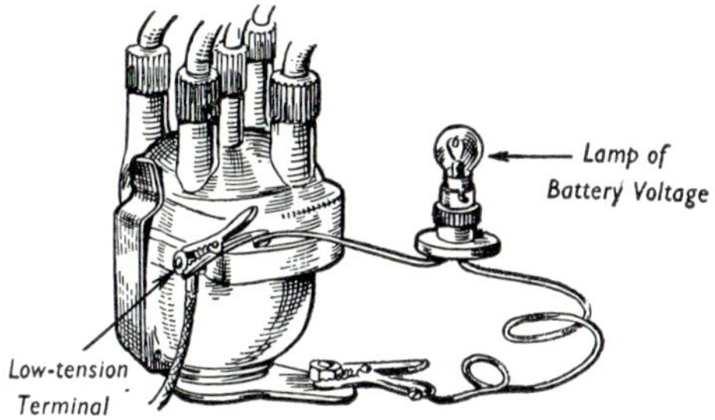

FIG. 44. TIMING THE IGNITION WITH THE AID OF AN INDICATOR LIGHT

out only with the aid of special equipment furnished by the manufacturer.

Automatic Timing Controls. Most distributors nowadays are fitted with a centrifugal timing control consisting of two pivoted weights built into the lower part of the distributor body. These are restrained by governor springs, and fly apart as the engine speed increases, rotating the sleeve which carries the cam and rotor, and so advancing the ignition timing. If, owing to rust, the weights bind on their pivots or the base plate, it will be necessary to remove the contact-breaker plate in order to dismantle and clean them. They should be lubricated with thin oil.

If one restraining spring appears to be slack, it should not be assumed that it is faulty; this is intentional in some distributors in order to provide a rapid initial advance with only one spring in action, and a slower rate of advance over the remainder of the speed range as the second spring becomes effective.

Often, a suction-operated control is fitted in addition. This operates by means of the depression in the induction system.

and is connected by a pipe line to a drilling in the carburettor on the atmospheric side of the throttle. Thus, when the engine is idling with the throttle closed no suction exists in the pipe, and the ignition is fully retarded. On opening the throttle during acceleration, when the engine is pulling hard, there is still little depression present in the vacuum timing control; although the centrifugal control will be advancing the ignition progressively, the degree of advance will not be sufficient to

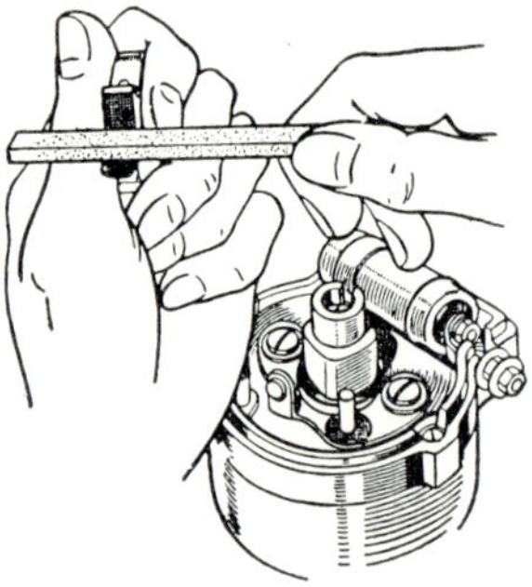

Fig. 45. Pitted Contact Breaker Points may be Trued Up with a Fine Carborundum Stone, Providing that the Surfaces are Kept Square

cause pinking. If the throttle is partly closed at cruising speeds, however, the vacuum control comes into action and advances the ignition still further, thus giving maximum economy.

Using a Neon Timing Light. It is difficult to test the action of automatic timing controls under actual running conditions without the use of a neon timing light. This consists of a bulb filled with neon or other gas, and connected to a sparking plug. As the plug fires the bulb flashes, and the rapid sequence of flashes provides a stroboscopic effect in which any part of the engine which is operating at crankshaft speed, or a definite multiple of crankshaft speed, appears to be stationary.

If, therefore, the flywheel is viewed through the inspection aperture with a timing light of this type, the timing mark will appear to be stationary at a steady engine speed, and the degree of advance can immediately be judged by its position in relation to the indicator line on the clutch housing or crankcase.

As the engine is speeded up, the timing mark will appear to move in the opposite direction to the rotation of the flywheel; when a vacuum timing control is fitted, the mark should move still farther backwards when the throttle is partly closed, and should move forwards, indicating that the ignition is being retarded, immediately the throttle is opened, and before the

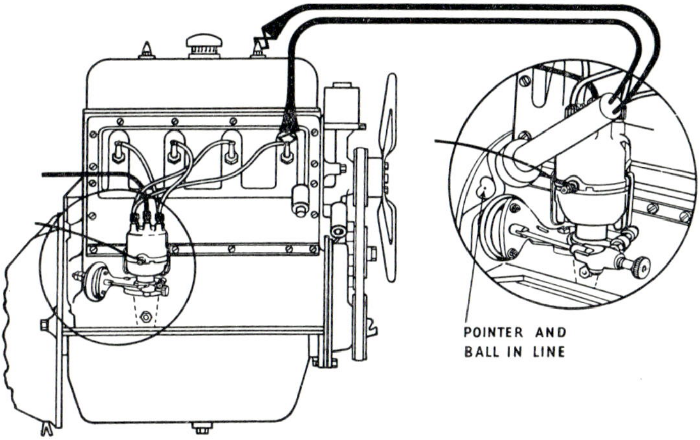

FIG. 46. USING A NEON TIMING LIGHT (STROBOSCOPE) TO CHECK THE IGNITION TIMING

engine picks up sufficient speed to cause the centrifugal control further to advance the ignition.

As a makeshift aid to fault-tracing, effective results can be obtained in a darkened garage by wiring a sparking plug in series with one sparking plug lead and using the flash of the spark at the points to illuminate the flywheel. The timing mark on the flywheel will more readily be visible if it is picked out with white or aluminium paint. On some engines, in fact, a polished steel ball is pressed into the flywheel rim to enable tests with timing lights to be carried out easily.

THE SPARKING PLUGS

Chart No. 7, based on one prepared by the A.C. Sparking Plug Company, summarizes the various sparking plug troubles

Normal Plug Condition
The insulator should be coloured from straw to rich brown. Carbon deposit should be light

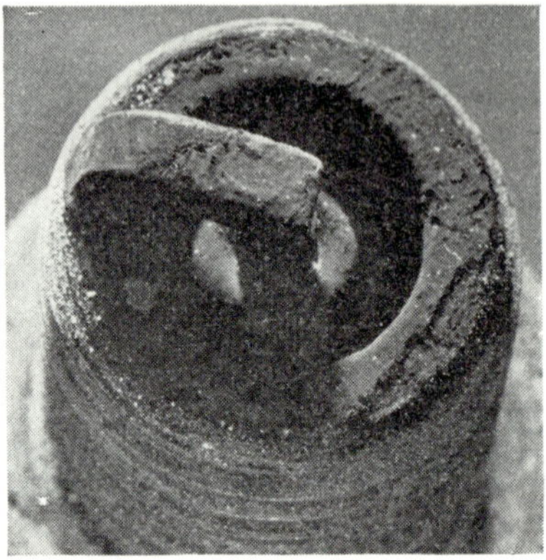

A "Sooted" Plug
This is a sign of an overrich mixture, possibly combined with a plug that is too cool for for the engine

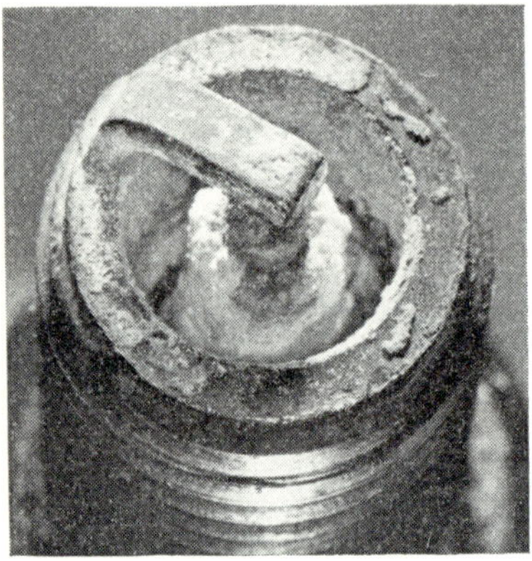

Evidence of Excessive Heat
The insulator is bleached and blistered. The plug must be scrapped

An "Oiled" Plug
This condition indicates either an excess of oil in the combustion chamber or too "cold" a plug, or possibly a combination of both faults

Fig. 47. Choosing a Suitable Grade of Sparking Plug

likely to be encountered. In amplification it is necessary only to draw attention to one or two common fallacies which sometimes lead even expert mechanics astray when diagnosing faults.

It is often taken for granted, for instance, that a deposit of dry black soot on a plug is necessarily caused by too rich a mixture. The fact is frequently overlooked that this condition can also be caused when a plug does not attain a high enough temperature to burn off the soot, which is an inevitable product of combustion of hydro-carbon fuels. The best test is to drive for some distance at full-throttle, and while still travelling at speed to put the gear lever into neutral and switch off the engine, keeping the throttle wide open until the engine has stopped. The condition of the internal insulators of the plugs will then give a good indication of the conditions prevailing at the time that the engine was switched off. The insulators should appear clean, and of a light biscuit-colour. If they are still sooty, it is likely that a grade of plug which attains a higher temperature will prove more satisfactory.

Plug Testing Fallacies. A number of misconceptions arise regarding the methods of testing the efficiency of a sparking plug. Many owners and mechanics still believe implicitly that if a spark jumps the plug points when the plug is laid on the cylinder head, and the engine is rotated, the plug must be satisfactory. Firing under these conditions, however, is a matter very different from firing when subject to a compression of upwards of 100 lb. per square inch in the combustion chamber. As already emphasized, the electrical resistance across the plug points is increased to such an extent under these conditions that the current may jump across the internal or external insulator instead of across the points. The only effective tests, therefore, is one carried out with the plug subjected to pressure, as in a plug-cleaning and testing machine of the type now widely used by garages. Even this test does not rule out the possibility of the current " tracking " across the insulation of the distributor cap.

Nor should too much faith be placed in the conventional types of neon plug-tester, used by so many owners. As a proof that a flash in the tester does not necessarily mean that a spark occurs at the plug points, a plug lead should be detached and held clear of any metal surface. On applying

CHART No. 7

SPARKING PLUG DEFECTS AND REMEDIES

Condition of Plugs	Remedy	Further Checks and Precautions
Gap incorrect. Electrodes badly worn or burned away.	Set gap to manufacturer's recommendation. If plug in service for 10,000 miles or more, replace. If electrodes badly burned in only five or six thousand miles, replace with cooler plug.	Check ignition system and battery. Check carburettor adjustment. Check timing. Set distributor points to manufacturer's recommendation.
External insulator broken.	Install new plug. Such breakage can be caused only through carelessness.	Use care in removing and installing plugs. Use a sparking plug spanner of the proper size, and keep it square.
Internal insulator cracked or broken at lower end.	Install a new plug. If broken plug is correct type, engine is probably being operated under hotter than normal conditions, and "cooler" plug should be installed.	This trouble may be caused by carelessness in re-setting plug gaps. Make all such adjustments by bending the side electrode only.
Upper part of external insulator blackened just above the shell, indicating leakage. Internal insulator not sooty.	Type of plug used is much too hot—install a cooler type. On detachable plugs gland nut is loose or washer faulty.	Gland nuts on detachable plugs should be tightened with box spanner. Use of set spanner or gripping in vice will distort nut.

Internal insulator blistered or "glassy." Insulator shows deposits of reddish or brownish colour.	Clean plugs with sparking plug cleaner: reinstall plugs and test. If condition is habitual, it indicates "too hot" operation; change to cooler type of plug.	Check timing. Check carburettor adjustment. Check for leaks in intake manifold and in cooling system. Fuel used may be unsuitable.
Internal insulator covered with dry, black soot.	Clean plug. Set gap to manufacturer's recommendation. Reinstall and test. If sooting is habitual, replace with hotter type.	Check for too-rich carburettor adjustment. Set distributor points to manufacturer's recommendation. Use choke sparingly; see that it is not sticking.
Internal insulator caked with oily carbon or soot.	Check for correct type. Clean plug, set gap to manufacturer's recommendation. Reinstall plug and test. If condition is habitual or plug of incorrect type, replace with hotter plug.	Check for too-rich carburettor adjustment. Set distributor points to manufacturer's recommendation. Use choke sparingly; see that it is not sticking. Check for too much oil in crankcase. Check ignition system and battery. Check timing. Check for leaky or sticking valves. Check for worn or loose pistons and piston rings.
Plug oily, but not sooty or carbonized—spark gap filled with oil.	Swill with petrol and clean. Set gap to manufacturer's recommendation. Examine plug carefully for cracked insulator. Reinstall plug and test. If plug does not operate satisfactorily, the trouble is elsewhere.	Check ignition system and battery. Set distributor gap to manufacturer's recommendation. Check timing. Check for too rich mixture. Check for excessive use of choke. Check choke for sticking. Check for too much oil in crankcase. Check for worn pistons or rings. Check for leaky or sticking valves.

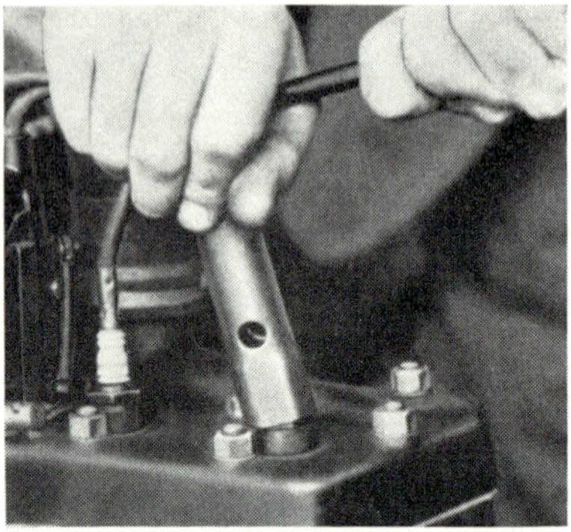

Fig. 48. How Sparking Plug Insulators are Broken. The Spanner Must be Kept Squarely on the Plug and Equal Pressure Applied with Both Hands

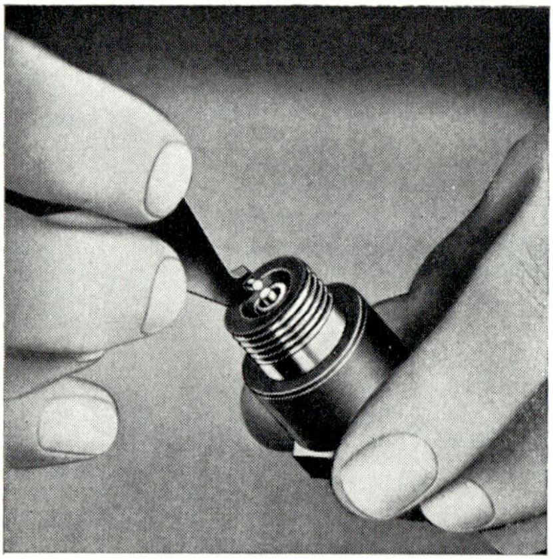

Fig. 49. The Sparking Plug Gap Should Be Adjusted by Bending the Side Electrode, Preferably Using a Special Tool, as Shown

the neon tester to the end of the lead, regular flashing will occur.

The only condition, in fact, which will prevent the tester from flashing is complete short-circuiting of the current caused by the sparking plug points or insulator being bridged by oil, moisture, or carbon. To diagnose plug condition by studying the intensity of the flash in the tester, as opposed to the presence or absence of a flash, calls for considerable experience.

A further example of faulty reasoning is the owner who attempts to cure misfiring by tightening each sparking plug terminal, overlooking the fact that a tight contact here, or even *actual* contact, is unnecessary. Before reaching the plug leads, the current jumps an appreciable airgap between the distributor rotor and the electrodes in the distributor-cap. The importance of really clean and tight connexions throughout the low tension circuit, on the other hand, is often overlooked. It is not unusual to find that the coil and contact breaker terminals are loose, corroded, or coated with greasy dust. Yet, in a six- or twelve-volt circuit, even a slight voltage drop caused by poor connexions will have an appreciable effect on ignition efficiency.

CHAPTER V

DIAGNOSING CARBURETTOR AND FUEL SYSTEM FAULTS

WITH the aid of Charts Nos. 8 and 9 it should be possible to diagnose most carburettor and fuel-supply system faults. Carburation is such a wide subject, however, and many modern carburettors are such complex instruments, that the novice will be well advised to confine himself to the detection and remedying of obvious faults, leaving to the expert more advanced work, such as stripping and overhauling the carburettor and tuning it to give characteristics different from standard.

It is worth remembering, however, that the carburettor manufacturers can generally supply service booklets and data charts dealing with any particular instrument, and are usually only too willing to give advice and assistance regarding their products. This chapter, therefore, will be devoted to a general survey of elementary fault tracing and diagnosis.

Diagnosing Starting Carburettor Defects. On many British carburettors the rich mixture required for starting is supplied by a small auxiliary starting carburettor, which is built into the main instrument, but which operates independently of the normal idling, acceleration and power systems.

When discussing cold-starting problems in Chapter II, it was emphasized that the starting control must be correctly set and that the throttle must be fully closed when the starting carburettor is in use, since the starting carburettor feeds through a port on the engine side of the throttle, and depends for its suction on the depression present in the induction manifold when the butterfly valve is practically closed at the normal idling setting.

On some carburettors the supply of the starting mixture is controlled by a valve. If the valve faces become worn or scored, a rich mixture will be drawn into the engine when the dashboard control is in the " closed " position, causing increased fuel consumption. As a check, remove the petrol jet

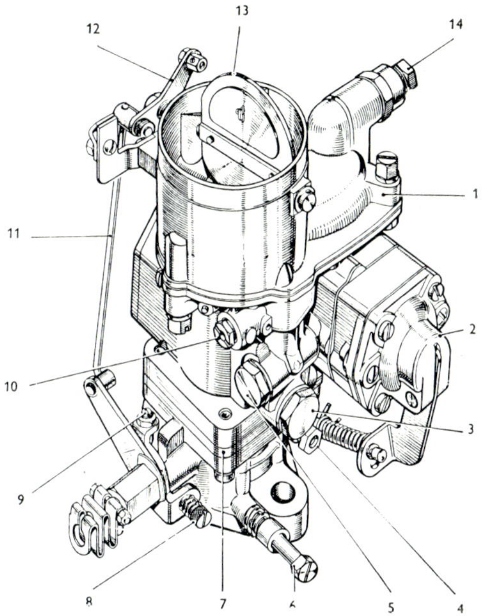

FIG. 50. SERVICING POINTS ON A TYPICAL SOLEX CARBURETTOR

1. Float chamber cover
2. Acceleration pump
3. Main-jet holder
4. Vacuum-timing connexion
5. Acceleration pump jet
6. Slow-running mixture adjustment
7. Heat-insulating washer
8. Slow-running speed adjustment
9. Fully-open throttle stop
10. Slow-running jet
11. Throttle-strangler (choke) interconnexion rod
12. Strangler (choke) lever
13. Strangler (choke) valve
14. Fuel pipe union

and block the holes with a matchstick. If, after a road test, the petrol consumption is improved, a new valve should be fitted.

On some other designs the starting mixture control is in the form of a spring-loaded plunger. Obviously, if the

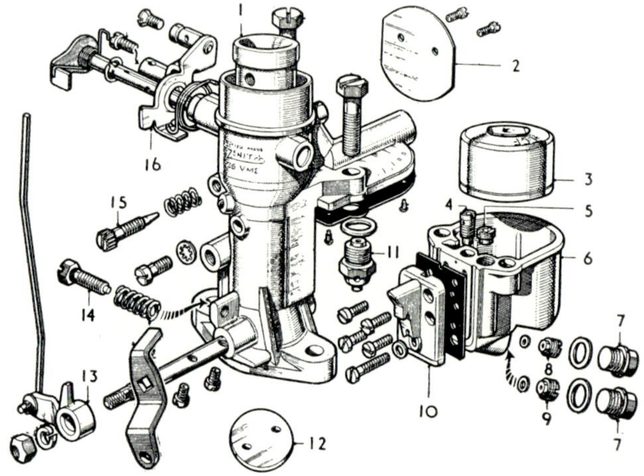

FIG. 51. IN THIS ZENITH CARBURETTOR THE MAIN AND COMPENSATING JETS ARE RETAINED BY EXTERNAL PLUGS AND THE THROTTLE AND STRANGLER ARE INTERCONNECTED FOR STARTING

1. Choke tube
2. Air strangler
3. Float
4. Slow-running jet
5. Capacity-well screw
6. Float chamber
7. Jet plug
8. Main jet
9. Compensating jet
10. Emulsion block
11. Needle-valve assembly
12. Throttle disk
13. Throttle interconnecting lever
14. Throttle stop screw
15. Air-regulating screw
16. Strangler-operating lever

plunger sticks, the starting mixture will be drawn into the engine during normal operation. Too weak a starting mixture will be supplied, on the other hand, if the small spring-loaded air valve should stick in the open position. This valve should remain on its seating until the engine has fired, when the additional depression in the suction tube and passage leading to the valve is sufficient to overcome the resistance of the spring. On the majority of modern carburettors, however, a spring-loaded air strangler flap is used, as described on page 99.

Faulty Thermostatic Starting Carburettor Controls. On some cars the starting carburettor is brought into action, and cut out, automatically, by a thermostatic device operated by the heat of the engine.

On the Solex Thermostarter three faults are possible. Each results in the starting carburettor remaining in action when the

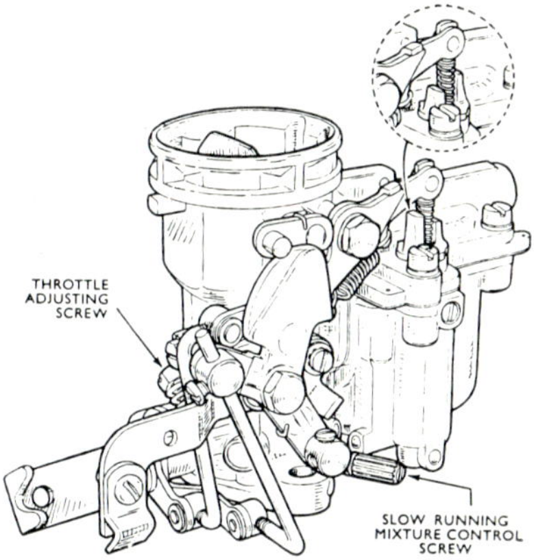

FIG. 52. IDLING SPEED AND MIXTURE ADJUSTMENT ON A ZENITH CARBURETTOR

Inset: adjustment of acceleration pump stroke by stops of different heights beneath lever

engine has warmed up : a distorted or incorrectly adjusted bi-metal strip in the control box attached to the exhaust manifold ; air leaks in the pipe between the thermostatic control and the membrane or diaphragm housing on the carburettor ; a cracked membrane ; or air leakage at the diaphragm chamber flanges.

To diagnose any of these faults, disconnect the vacuum pipe from the membrane chamber while the engine is idling and seal the aperture in the housing with the finger. If the starting carburettor goes out of action, the fault lies in the thermostat or pipe line ; if it does not, the diaphragm is cracked or air is leaking at the flanges.

On the S.U. thermostatically-controlled starting carburettor the starting mixture is controlled by a disk-valve which is

CHART No. 8

CARBURETTOR FAULTS

In this chart only carburation and associated faults are dealt with : it should be used, therefore, when starting troubles, misfiring, and loss of power have been definitely attributed to the carburettor, after reference to Charts Nos. I, II, III and IV.

STARTING DIFFICULT OR IMPOSSIBLE

- ENGINE COLD — Check setting of starting control (see text) — Check for delivery of fuel as in Chart No. 1
  - Petrol available in float chamber — Examine starting carburettor petrol and air jets — Jets clean
    - Air or petrol passages in starting carburettor obstructed.
    - Incorrect starting carburettor petrol or air jet.
    - Automatic air valve in carburettor faulty.
    - Throttle not fully closed.
    - Automatic choke faulty.
    - Stale petrol in float chamber.
    - Jet piston sticking (S.U. and Stromberg CD carburettors).
    - Starting carburettor thermostat faulty or incorrectly set.
    - Air leaks in induction system.
    - Manifold drain valve stuck open or jet omitted.
  - No petrol in float chamber but pump delivers fuel — Examine carburettor filter — If clean
    - Carburettor needle valve stuck.
  - Pump does not deliver fuel
    - Check for fuel supply system faults as in Chart No. 9.
- ENGINE WARM OR HOT — Check petrol supply as above — If petrol is available — Examine carburettor and manifold drain pipe for signs of flooding
  - Evidence of flooding or leakage from drain pipe
    - Excessive use of choke or starting carburettor should be unnecessary when engine is warm or hot.
    - Carburettor needle valve struck.
    - Petrol level too high.
    - Choke or mixture control remaining in operation.
    - Pumping of accelerator pedal causes acceleration pump to deliver excess fuel.
  - No flooding or leakage
    - Starting carburettor troubles (see above).
    - Weak mixture or air leaks (see above).

- **Poor Slow Running or Engine Stalls** — Check idling mixture setting and note whether engine "hunts" when idling with emission of black smoke from exhaust (see text)
  - Engine "hunts": **exhaust has sweet odour: mixture too rich**
    - Incorrect setting of idling mixture control.
    - Carburettor flooding.
    - Float chamber petrol level too high.
    - Fuel pump pressure excessive.
    - Starting device not cutting out.
    - Fuel drawn through autovac suction pipe due to faulty vacuum valve.
  - Engine runs irregularly and tends to die out—**exhaust sounds "splashy"** and has no odour; **mixture too weak**
    - Incorrect setting of idling mixture adjustment.
    - Idling jet obstructed.
    - Petrol starvation (see above)

- **Lack of Power**
  - **When Accelerating** — Examine and clean air cleaner — If clean, check petrol level and fuel supply — If correct, check mixture strength by sparking plug appearance (see text) or with exhaust gas analyser
    - Mixture too weak, probably accompanied by spitting back in carburettor
      - Fuel starvation (see above)
      - Water or sediment in carburettor.
      - Compensating jet choked.
      - Acceleration pump not operating or incorrectly adjusted.
      - Hydraulic damper needs topping-up (S.U. and Stromberg CD carburettors).
      - Jet piston sticking (S.U. and Stromberg CD carburettors).
      - Wrong size of jet or choke tube fitted.
      - Manifold heat control does not give high enough temperature.
      - Multiple carburettors not correctly synchronized.
    - **Mixture too ricb**
      - Wrong size of compensating jet.
      - Acceleration pump stroke too long.
      - Starting carburettor or strangler remaining in operation.
      - Manifold heat control maintains too high a temperature.
      - Multiple carburettors not correctly synchronized.

CHART No. 8 (continued)

- **At High Speeds**
  - Examine and clean air cleaner
    - If clean, check petrol level and fuel supply
      - If ample fuel available, check mixture strength (see above)
        - Mixture too weak
          - Main or power jet too small
          - Jets obstructed.
          - Automatic air or petrol valve faulty (see text).
          - Throttle not opening fully.
          - Multiple carburettors not correctly synchronized.
        - Mixture too rich
          - Main or power jet too large.
          - Economy device not operating (see text).
          - Multiple carburettors not correctly synchronized.
      - If symptoms of fuel starvation
        - Vapour lock in fuel line.
        - Fuel system faults (see Chart No. 9).

- **Excessive Fuel Consumption**
  - Examine carburettor and fuel supply system for leakage of petrol.
    - Clean air filter
      - Check idling mixture adjustment
        - Verify petrol level in float chamber
          - If all foregoing points are correct
            - Starting device or choke remaining in action too long.
            - Acceleration pump stroke too long.
            - Economy device not functioning.
            - Jets too large or reamered out.
            - Worn needle or jet (S.U. and Stromberg CD carburettors).
              Sticking piston (S.U. and Stromberg CD carburettors).
            - Petrol pump pressure too high.
            - Manifold heat control incorrectly set.
            - Carburettor worn.
            - Multiple carburettors no correctly synchronized.

raised from its seating by an electro-magnet, the windings of which are energized by current flowing from the ignition switch through the windings to a thermostatic switch which is influenced by the heat of the engine cooling water. If the contact points are dirty, if the bi-metal strip is incorrectly adjusted, or if the electrical connexion between the switch and the carburettor is faulty, the starting device will not operate. Test the thermostatic switch by short-circuiting the terminal at its centre to the switch body. This should bring the starting carburettor into action. If it does not, check the wires and connexions between the ignition switch and the carburettor, and between the carburettor and the thermostatic switch. If these are in order, the fault must lie in the electro-magnet ; alternatively, the valve plunger is sticking.

If the starting mixture remains in action too long, check the thermostatic switch adjustment and the disk valve, which may be held off its seating by a particle of grit. The starting mixture strength is controlled by the collar stop screw adjacent to the electro-magnet ; screwing it downwards weakens the mixture.

Automatic Choke Controls, The simple strangler or choke control which shuts off most of the air supply to the carburettor when starting, thus inducing a very rich mixture, is still used, with certain refinements, on a number of carburettors. In some designs the strangler valve is offset on its spindle and is spring-loaded, so that the additional suction created when the engine fires opens the valve to a degree which depends on the engine speed. In other designs a spring-loaded flap in the valve is similarly opened. Fault-tracing is thus confined to examination of the valve or flap for freedom of operation, and the spring for strength.

An additional refinement is to couple the strangler to a thermostatic spring in a box attached to the exhaust manifold. The heat of the exhaust when the engine is running causes the spring to unwind and rotate a lever which opens the strangler valve through the medium of the coupling rod. Points to check on this type of control are the correct setting of the coupling rod, which may be adjustable for length, and the operation of the thermostatic spring, which may also be adjustable.

Thermostatic or manual control of the strangler may also

be combined with a mechanical linkage between the strangler control and the throttle lever. This gives the correct throttle opening for starting from cold, a fast idling speed when warming up, and normal idling speed when the engine has attained its running temperature, and is effected by a three-step cam in the linkage. Correct adjustment of the linkage in

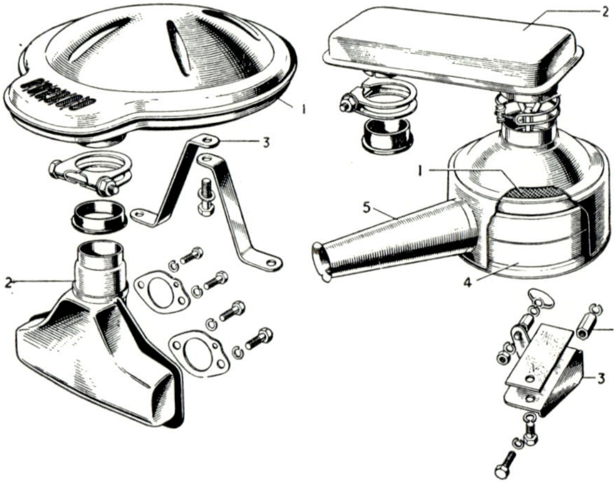

FIG. 53. TYPICAL AIR FILTERS

1. Filter element
2. Connexion to carburettor
3. Mounting bracket
4. Oil bath
5. Air-intake horn

accordance with the manufacturer's instructions is therefore essential.

Checking the Idling Adjustment. The only adjustments normally required by a carburettor are the correct setting of the idling mixture and speed adjustments. The mixture strength is generally adjusted by means of a spring-loaded screw which regulates the amount of petrol or the amount of air supplied. The idling speed is controlled by an adjusting screw which acts as a stop for the throttle as it approaches the fully closed position.

The idling mixture is best adjusted with the aid of a vacuum gauge, the mixture-regulating and idling-speed controls being adjusted alternately until the highest reading is recorded

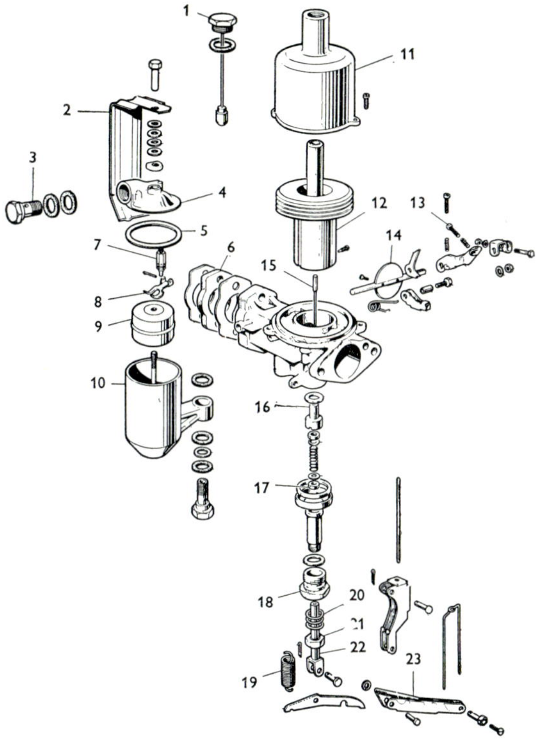

FIG. 54. TYPICAL S.U. CARBURETTOR, WITH SOME POSSIBLE FAULTS

1. Damper piston ineffective
2. Mounting bracket loose or bent
3. Petrol union leaking
4. Float chamber cover distorted or loose
5. Gasket cracked
6. Heat-insulating washer not fitted. Air leaks at joints
7. Faulty needle valve
8. Valve-operating fork incorrectly adjusted
9. Punctured or dented float
10. Sediment in float chamber
11. Piston chamber dirty, worn or distorted
12. Piston dirty or worn
13. Throttle-stop screw incorrectly set
14. Worn throttle disk, spindle and bearings
15. Needle wrong size, bent or incorrectly set
16. Incorrectly centred jet bearing
17. Leaking jet seal
18. Loose jet retaining nut
19. Jet lever spring broken or unhooked
20. Jet-adjusting nut spring missing
21. Jet-adjusting nut incorrectly set
22. Jet worn
23. Jet-control lever incorrectly adjusted

by the gauge. It will often be found that the reading does not alter appreciably if the mixture is slightly enriched or weakened, once the approximately correct setting has been found. It is best in these cases to use the richest setting which can be employed without causing the gauge reading to drop.

If a vacuum gauge is not available, too rich a mixture can be detected by the engine running "lumpily," while the

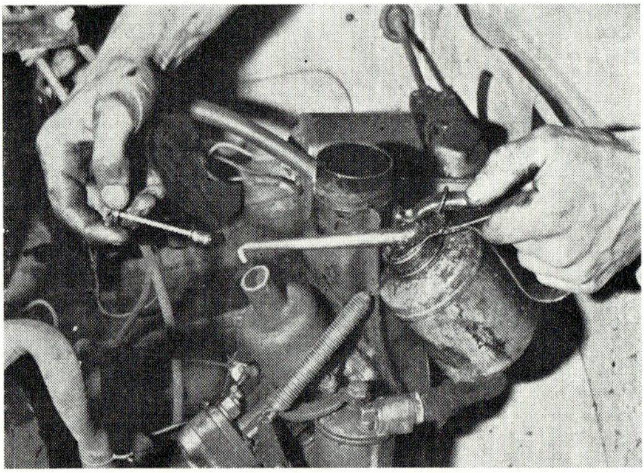

Fig. 55. Topping-up the Damper-piston Chamber on an S.U. Carburettor

exhaust gas will have a sweet odour. There may be signs of black smoke from the tail pipe. Too rich a mixture will cause the exhaust to sound "splashy," and the engine may miss several beats at intervals. The gas will be odourless. On suddenly opening the throttle wide, the engine may hesitate for a moment when the mixture is too rich, but will then accelerate rapidly. If the mixture is too weak, the engine will hesitate, spit back in the carburettor, and may stall.

The adjustment of idling-mixture strength on S.U. and Zenith-Stromberg carburettors (*see* Figs. 54 and 57) is carried out by means of a jet-adjusting stop nut, or by a mixture control screw. The mixture strength is determined by the position of the tapered jet needle in the jet. The needle is raised by a piston as engine speed increases, thus withdrawing the needle from the jet and enlarging the orifice through which the petrol enters the air stream.

When the engine is idling, however, the piston is at its lowest position. To alter the mixture it is therefore necessary to move the jet upwards or downwards. If the jet is raised, the mixture will obviously be weakened, while lowering the jet will enrich the mixture. The mixture control rod or wire,

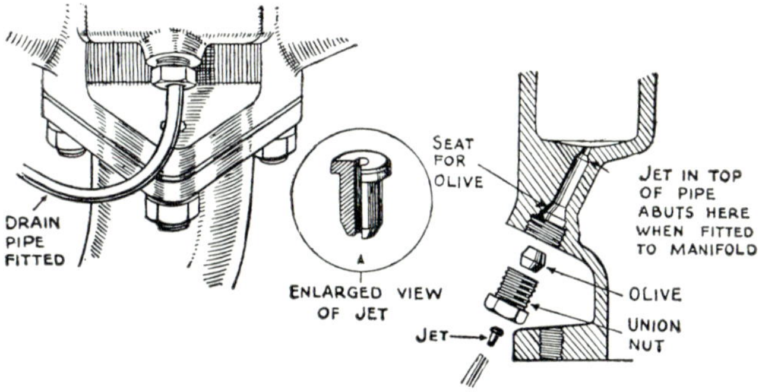

FIG. 56. OVER-RICH MIXTURE CAN BE CAUSED BY THE MANIFOLD DRAIN PIPE OR JET BECOMING CHOKED. A WEAK MIXTURE RESULTS IF THE JET IS OMITTED, AS ON THIS HILLMAN EXAMPLE

when fitted, should be disconnected from the lever which actuates the jet, and the flange of the jet should be kept in contact with the adjusting nut during adjustment. The method of judging the correct mixture is as already outlined.

Tuning Carburettors for Power and Economy. Owing to the very wide differences in the manner in which carburettor manufacturers have tackled the problem of providing the maximum power with the utmost economy, the only advice which can be given regarding tuning a carburettor is that the car-owner should obtain from the manufacturers a list of alternative jet settings, choke tube sizes, economy device and acceleration pump adjustments. It must be emphasized that economy can be obtained only at the expense of power or maximum speed and vice versa. The standard setting of the carburettor represents the best compromise between these conflicting factors, and should be altered only in exceptional cases.

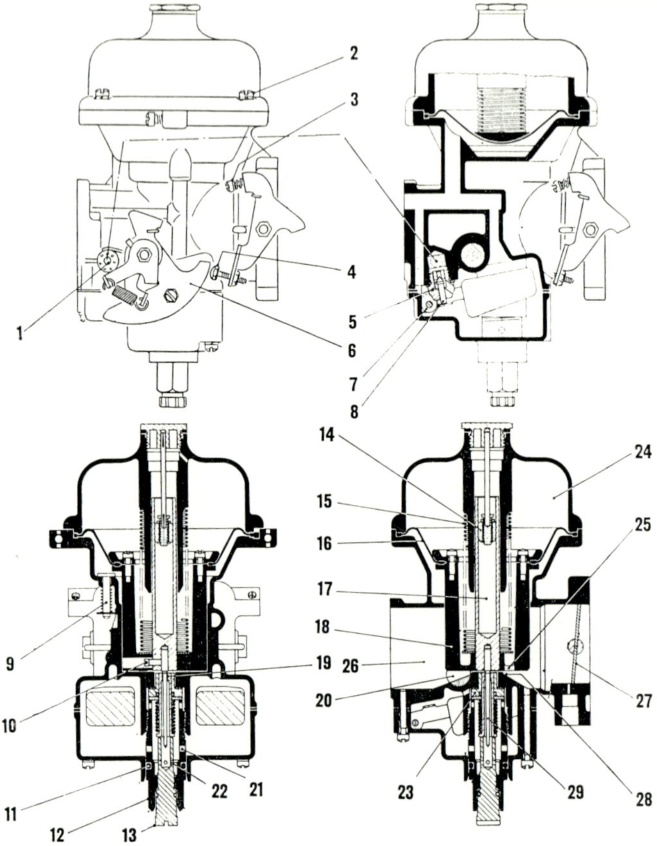

Fig. 57. Possible Faults in a Zenith-Stromberg Carburettor

1. Leakage at petrol inlet
2. Screws loose
3. Throttle-stop screw incorrectly adjusted
4. Fast-idle screw incorrectly adjusted
5. Needle seating worn
6. Fast-idle cam bent
7. Float arm incorrectly set
8. Needle valve leaking
9. Spring-loaded pin binding
10. Needle locking screw loose
11. "O" ring leaking
12. Jet assembly incorrectly centred
13. Jet adjusting screw incorrectly set
14. Damper oil level incorrect or rod bent
15. Coil spring weak
16. Diaphragm defective
17. Guide rod binding
18. Air valve dirty
19. Jet orifice choked
20. Starter bar not operating correctly
21. Inlet hole choked (unlikely)
22. Inlet hole choked
23. Jet bush incorrectly centred
24. Loss of vacuum in air chamber
25. Air-valve drilling blocked
26. Bore of carburettor incorrect for engine
27. Throttle loose on spindle
28. Bridge damaged
29. Jet metering needle wrong size, bent or incorrectly set

In practice, adequate improvement in power or economy can generally be obtained by careful checking of the various points listed under the appropriate headings in Chart No. 8.

It must also be realized that engine and ignition faults may in fact be responsible for deficiencies blamed in error on the carburettor.

It is difficult, moreover, to arrive at a true assessment of any alteration in jet size or other adjustment without the use of an accurately calibrated flow-meter to measure petrol consumption, which must be tested over a closed road circuit, the greatest care being taken to employ exactly the same speeds and driving methods during each test. As an alternative, tuning can be carried out with the aid of an exhaust gas analyser.

Poor Acceleration. Two points which are covered under the heading "Lack of Power" in Chart No. 8 are, perhaps, worth amplifying. In the first case, Zenith-Stromberg and modern S.U. carburettors incorporate a small damper piston, operating in an oil-filled chamber in the upper end of the piston rod (*see* Figs. 54, 55, and 57), which prevents the piston rising too quickly when the throttle is opened suddenly. If the chamber is not kept topped up with the correct grade of engine oil, or if thin machine oil is used in error, the damping action will be inadequate and momentary starvation will occur when the throttle is snapped open, causing a "flat spot" and, possibly, spitting-back in the carburettor.

Other types of carburettor often include an acceleration pump which provides temporary enrichment of the mixture during the acceleration phase. In many cases the pump is operated by linkage which is connected to the throttle lever on the carburettor and it is possible to adjust the stroke of the pump: there may be alternative holes in the lever or alternative stops may be provided beneath the lever, as shown in Fig. 52. It is thus possible to select a longer stroke of the pump piston during cold weather, when additional enrichment is desirable. If, however, the more economical "summer" setting is retained during the winter, hesitant pick-up will probably be experienced.

Synchronizing Multiple Carburettors. On engines fitted with two or three carburettors, tuning is often complicated by the necessity to synchronize the carburettors so that each is exactly "in step" with the others at every throttle

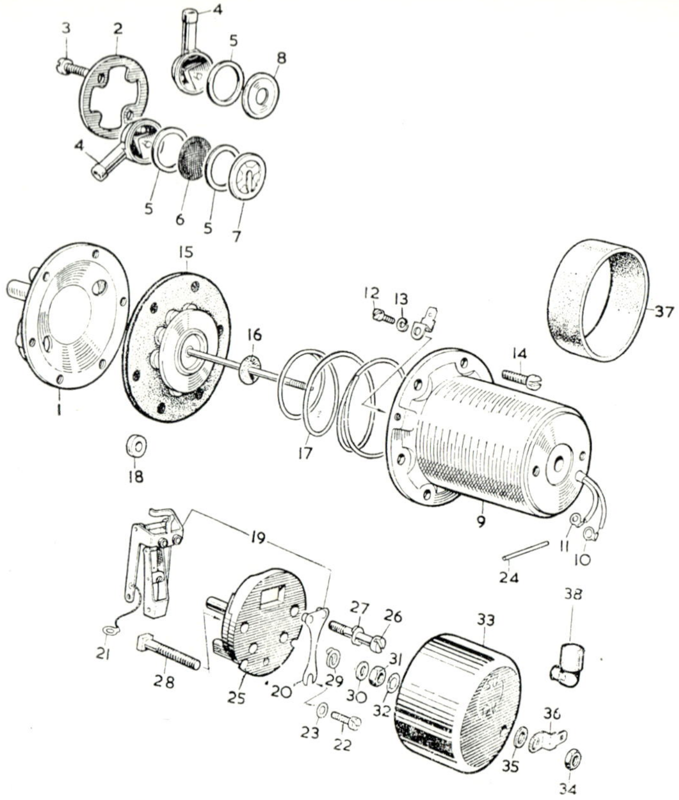

Fig. 58. S.U. AUF 200-type Fuel Pump

Other S.U. models are similar, except for the arrangement of the valves

1. Body
2. Spring clamp plate
3. Screw
4. Inlet/outlet nozzle
5. Sealing washer
6. Filter
7. Inlet valve
8. Outlet valve
9. Coil housing
10. 5 B.A. terminal tag
11. 2 B.A. terminal tag
12. Earth screw
13. Spring washer
14. Screw, housing to body
15. Diaphragm assembly
16. Impact washer
17. Spring
18. Roller
19. Rocker and blade
20. Blade
21. 2 B.A. terminal tag
22. Screw for blade
23. Dished washer
24. Spindle for contact breaker
25. Pedestal
26. Pedestal to housing screw
27. Spring washer
28. Screw for terminal
29. Spring washer
30. Lead washer for screw
31. Nut for screw
32. Spacer—nut to cover
33. End cover
34. Nut for cover
35. Shakeproof washer
36. Lucar connector
37. Sealing ring
38. Non-return valve

opening. Here again it is wise to consult the manufacturer of the car so that advantage may be taken of his experience of the best method of solving the problem on a particular engine. Not only does tuning involve obtaining the correct mixture strength; it also involves obtaining the right throttle opening at

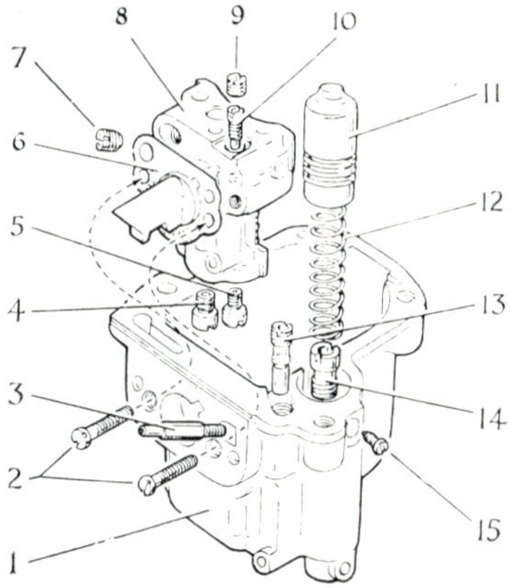

FIG. 59. JETS AND ACCELERATION PUMP COMPONENTS OF A ZENITH CARBURETTOR

1. Float chamber
2. Emulsion block screws
3. Acceleration pump jet
4. Main jet
5. Compensating jet
6. Emulsion block gasket
7. Capacity vent
8. Emulsion block
9. Slow-running air bleed
10. Slow-running jet
11. Acceleration pump piston
12. Acceleration pump piston return spring
13. Acceleration pump delivery ball valve
14. Acceleration pump suction ball valve
15. Acceleration pump piston retaining screw

various engine speeds. Some experts are able to obtain good results by ear alone, listening to the hiss at the air-intake of each carburettor. Better still, however, is the use of a vacuum gauge, or a simple venturi device which is attached to each carburettor-intake in turn, the mixture and speed being adjusted in each case to give the same maximum reading on the gauge. Since the reading will depend on the diameter and angles of the venturi, engine characteristics, and other variable factors, it will be necessary to follow the instructions issued by the manufacturer of the gauge. Some trial-and-error tests may, in fact, be required before the most satisfactory results are obtained.

Fuel Supply System Faults

With the aid of Chart No. 9 it should be a fairly simple matter to trace a fuel pump fault. When a trouble has definitely been localized in an electric or mechanical pump there is a great deal to be said for taking advantage of the exchange service operated by most of the stockists or distributors of the particular type of pump, the defective pump being exchanged for a reconditioned unit at a price which covers the average cost of repair.

Simple faults, however, can often by put right without extensive dismantling : typical examples are given below.

S.U. Electric Pumps. Since the pump is electrically operated, complete failure is obviously most likely to be due to a disconnected lead or a bad contact at some point. The systematic tests given in the chart will lead to a choice of faults, among which a likely offender is too great a tension on the diaphragm of the pump. This will cause the contact points to remain open, owing to the fact that the tip of the inner rocker of the throw-over mechanism, illustrated at (19) Fig. 58, is not in contact with the baseplate of the mechanism. If the contact points are badly blackened, this is another indication of excessive diaphragm tension.

To release the diaphragm, remove the pump from the car and remove the six screws that secure the magnet housing to the body of the pump. The diaphragm will probably adhere to the flange of the magnet housing, from which it must be carefully separated with the blade of a penknife. The housing should be held with the diaphragm uppermost so that the eleven brass rollers in a groove beneath the diaphragm do not fall out of place. Do not rotate the diaphragm, as this will affect the adjustment of the throw-over gear.

Re-fit the magnet housing to the body of the pump, but do not tighten the screws until the hinge pin on which the rockers of the throw-over gear pivot has been withdrawn. This will allow the diaphragm return spring to extend the diaphragm fully. The screws can then be tightened, the rockers lined up, and the hinge pin replaced.

If the diaphragm is damaged or cracked it can be removed by rotating it to unscrew the armature rod from the inner rocker. The replacement diaphragm and armature rod are then fitted by reversing the rotation, slipping the rollers into place beneath the diaphragm during the process. Keeping the contact spring blade out of contact with the rocker, screw the armature rod inwards as far as possible, and then unscrew it slowly, pressing it inwards at intervals, until the throw-over mechanism operates correctly. Unscrew the armature by a further distance equal to four holes on the flange of the magnet housing to give the correct setting, and assemble and tension the diaphragm as just described.

Faulty Valves. If valve trouble has been diagnosed, inspection, cleaning and the installation of new valves, if necessary, is quite straightforward. On the L and HP pumps the valves are housed beneath the outlet union. Wash the discs and seatings in petrol. If they appear to be warped, renew them. If replacements are not available, the valves can be exchanged, since the seal of the delivery valve—the upper of the two, is less critical than that of the inlet valve.

On the dual pumps the valves are similarly arranged but are retained by a screwed plug above the outlet union. On the AUF 200 range of pumps, which are widely used on recent models, the valves take the form of melamine discs beneath the inlet and outlet unions, the latter being retained by a plate secured by two screws. There is a domed gauze filter beneath the inlet union, with a gasket on each side of it. When reassembling the valves, the tongue on the outlet valve cage should face outwards, and that on the inlet valve, inwards.

On the SP pump, the valves are also in the form of melamine discs but they are housed inside the base casting of the pump, which must be removed from the magnet housing in order to inspect them. The inlet valve is lightly loaded by a leaf spring and the outlet valve by a coil spring. Make sure, when reassembling, that the valves are replaced in the correct positions and with the outlet valve assembly the right way round, with the spring facing the interior of the base casting.

P.D-type S.U. Pump. The vertical, P.D. type of S.U. fuel pump, which is of the pressurizing, as opposed to the suction-feed, type, and is therefore fitted close to the petrol tank, is a sealed unit in which the pumping action of a solenoid-operated

CHART No. 9

FUEL SUPPLY SYSTEM FAULTS

- **S.U. Electric Pump**
  - **Does not Operate** — Check that current is reaching pump terminal — Verify soundness of earth connexion to pump or fit earthing wire — Examine contact points, which should be closed when pump is not operating
    - Points open
      - Diaphragm tension too great.
      - Armature rod incorrectly adjusted.
      - Armature rod binding.
    - Points closed
      - Points oily or badly pitted.
      - Faulty valves.
      - Faulty coil winding or connexions.
  - **Supplies Insufficient Fuel** — Examine and clean filter — Test with carburettor pipe disconnected from pump
    - Fuel not delivered copiously in rapid jets
      - Faulty suction or delivery valves.
      - Excessive diaphragm tension.
      - Armature rod sticking or incorrectly adjusted.
    - Fuel delivered satisfactorily
      - Carburettor filter choked.
      - Carburettor needle valve sticking.
      - Pipe line between pump and carburettor obstructed.
  - **Pump Buzzes or Beats Rapidly** — Examine carburettor or manifold drain pipe for flooding
    - Petrol flooding
      - Carburettor needle valve stuck or not seating correctly.
      - Carburettor float punctured.
    - No flooding
      - Air leak in suction line.
      - Vapour lock in fuel pipe.
  - **Pump does not Deliver Fuel** — Test with length of tubing attached to inlet union and dipping in can of petrol (see text)
    - Pump raises fuel at least 30 in.
      - Air leaks in suction line between pump and tank.
      - Obstructed suction pipe.
    - Pump does not raise fuel
      - Air leaks at filter bowl or cover joints.
      - Air leak at flange.
      - Faulty valves.
      - Insufficient diaphragm movement.
      - Worn linkage.
      - Fractured rocker arm.

- **Mechanical Pump**
  - **Pump does not Supply Sufficient Fuel** — Test with fuel pump analyser. (If not available, check for faults in last column)
    - Time taken to fill container exceeds one minute
      - Leaking gaskets.
      - Defective valves.
      - Worn linkage.
    - Air bubbles in fuel delivered
      - Air leaks in suction pipe (see above).
      - Loose filter bowl or cover.
      - Leaking valve plug.
      - Faulty diaphragm.
  - **Fuel Pressure Apparently too High, evidenced by Carburettor Flooding** — Test with analyser
    - Pressure correct for type of pump
      - Carburettor needle valve not seating correctly.
      - Carburettor float punctured
      - Petrol level in float chamber too high.
    - Pressure too high
      - Insufficiently flexed diaphragm.
      - Incorrect diaphragm return spring fitted.
      - Pump mounting gasket too thin—test with additional gasket between pump and engine mounting face.

- **Vacuum Fuel Supply System (Autovac)**
  - **Failure to Deliver Fuel** — Check all unions in suction pipe line for air leaks. Tighten top cover retaining screws — Inspect and clean filter in fuel inlet elbow — If foregoing points are in order, trouble must be due to:
    - Tank requires priming: pour petrol through delivery elbow orifice.
    - Grit on valve seatings.
    - Valve spring broken or over-stretched.
    - Float punctured: but most models have self-draining float.
    - Top cover gasket faulty.
    - Insufficient vacuum at full throttle.
    - Outer chamber dries up. Fit vacuum reserve tank or vacuum booster.

plunger is transmitted by low-viscosity oil to a thin terylene-film diaphragm, which forms one wall of the pumping chamber.

The pump cannot be dismantled, except for removal of the upper and lower covers, which will reveal, respectively, the contact points and the gauze filter. Clean the filter, check the condition of the cork washer and tighten the filter cover-plate

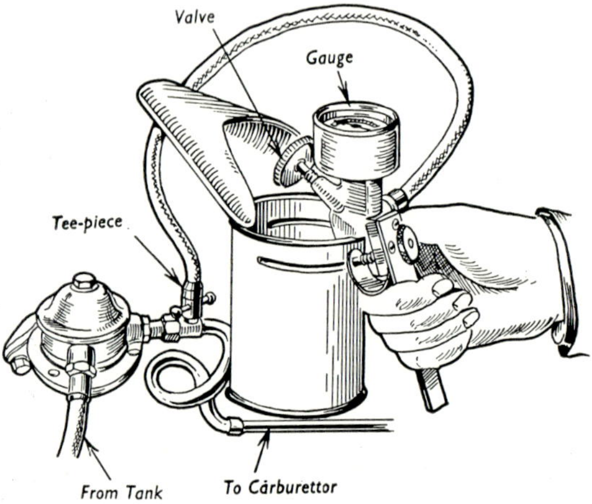

FIG. 60. THE A.C. FUEL PUMP ANALYSER

nut just sufficiently to prevent an air leak. If the contact points on the top plate of the pump are blackened, they may be cleaned by drawing a sheet of clean card between them while lightly pressing them together, but the greatest care must be taken not to distort or overstress the contact blades.

Finally, check that current is reaching the terminal on the feed wire and that the connexion between this wire and the upper terminal post on the pump is clean and sound. Check also the connexions on the earthing wire.

If, after these attentions, the pump fails to operate, it will be necessary to fit a replacement.

Diagnosing Mechanical Pump Troubles. The A.C. Fuel Pump Analyser is an invaluable aid to the diagnosis of mechanical fuel pump troubles, since tests can be carried out without removing the pump from the engine. When tests are made with the pump on the bench, the rocker arm being

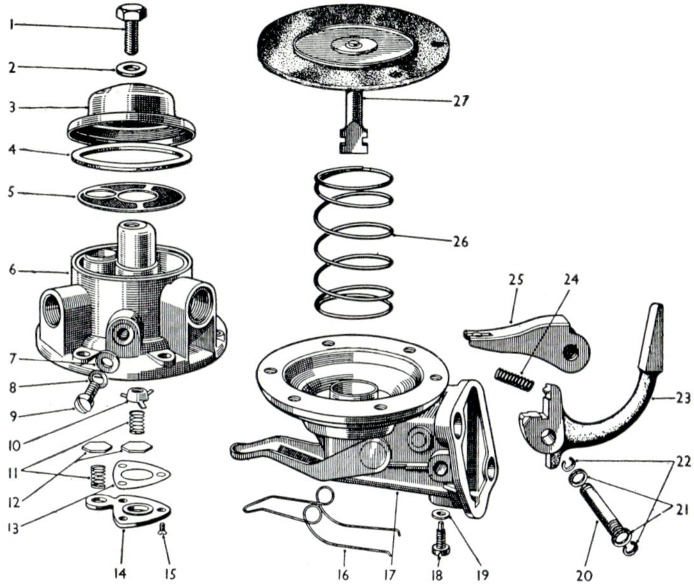

FIG. 61. AN "EXPLODED" SKETCH OF A MECHANICAL FUEL PUMP, SHOWING SOME POSSIBLE FAULTS

1. Top cover retaining-screw loose
2. Washer missing or defective
3. Top cover distorted
4. Cork washer defective
5. Dirty or perforated filter
6. Pump body cracked or otherwise damaged
7. Drain plug joint washer missing or defective
8. Drain plug washer missing or defective
9. Drain plug loose
10. Spring retainer bent or incorrectly assembled
11. Valve springs weak
12. Valves damaged, warped, or incorrectly fitted
13. Valve plate washer defective
14. Valve retainer plate damaged
15. Retainer plate screws loose
16. Priming lever spring weak or broken
17. Lower casting cracked or otherwise damaged
18. Rocker arm stop screw loose
19. Stop screw washer missing
20. Worn rocker arm pin
21. Rocker arm pin washers missing
22. Rocker arm pin clips insecure
23. Rocker arm bush and contact face worn
24. Rocker arm lever spring weak
25. Rocker link bent ; holes worn
26. Diaphragm spring tension incorrect
27. Diaphragm and pull rod assembly defective

operated by hand, the longer working stroke obtained sometimes tends to give a false idea of the efficiency of the pump.

The faults revealed by testing the pump with the analyser are tabulated in Chart No. IX. The correct pump operating pressure must be ascertained from the manufacturers of the car or of the pump, since it varies within fairly wide limits on different models. As a general rule it can be assumed that the pressure should be either $1\frac{3}{4}$, 2, $2\frac{1}{4}$, $2\frac{1}{2}$, $2\frac{3}{4}$, $3\frac{1}{4}$, $3\frac{1}{2}$, $3\frac{3}{4}$, 4, $4\frac{1}{2}$, $4\frac{3}{4}$, or, in rare cases, as high as 5 or $5\frac{3}{4}$ lb. per sq. in.

Testing without Analyser. After obvious sources of air or petrol leakage such as the filter bowl or cap gasket, flange washer, and loose unions, have been checked, the pump can be tested by disconnecting the inlet and delivery pipes and fitting a tube about three feet in length to the inlet of the pump. The end of the tube should dip into a can of petrol. If the pump is fitted with a hand-priming lever it can now be operated without removing it from the engine; if not, it must be detached so that the rocker arm can be worked by hand. The pump should be capable of drawing petrol when at least 30 inches above the level of the fuel in the can. If it fails to do so, one or other of those faults in the chart detailed under the heading "Pump does not Supply Sufficient Fuel" must be suspected.

Inspecting the Valves. If it is thought that the trouble is due to faulty valves, these may be inspected on some earlier types of pump by removing external valve plugs or, on another type, by first removing the filter disk and then unscrewing the valve plugs. Care must be taken not to lose the valve springs, and to refit the valves and springs in the same relationship to one another as they were before being dismantled. The polished side of the valve should face the seating.

On later types of pump the valves are in the form of caged assemblies, which must be renewed as complete units and are accessible only when the upper casting has been separated from the lower. Dismantling the pump for this purpose, or in order to renew the diaphragm, to check the diaphragm movement, or to renew any worn parts in the operating linkage, is not advisable, unless the special assembly and test stand is available, together with full instructions and a spare parts list covering the particular model of pump.

CHAPTER VI

COOLING SYSTEM DEFECTS

As will be seen from Chart No. 10, cooling system troubles can be broadly divided into three classes : overheating, over-cooling and loss of cooling water. The first and last groups of faults are naturally interdependent to a considerable extent.

Diagnosing Causes of Overheating. The most obvious cause of overheating, for instance, is lack of cooling water. This can be due to failure to top up the radiator at sufficiently frequent intervals to compensate for the normal loss of water which takes place owing to evaporation and leakage from the overflow pipe, as well as to less usual causes of leakage discussed later in this chapter.

A second elementary check is to verify that the fan belt is correctly tightened, especially as, when a pump is fitted, this generally drives the water pump as well.

As, on most modern engines, the dynamo is also driven by the belt, it will be evident that the belt is called upon to transmit considerable power—often amounting to several horsepower—while also absorbing the inertia loads of the fan and dynamo during sudden changes in engine speed. It is not always realized that, in addition, the fan belt acts as a form of crankshaft vibration damper when a torsional damper is not incorporated in the engine. The belt, therefore, has an exacting task to perform, and merits occasional attention.

Causes of Cooling Water Loss. The water level should always be checked when the engine is warm. As the water is heated, its volume increases, so that if the level in the header tank is restored when the engine is cold, water will be lost from the overflow pipe as the engine warms up unless the pipe discharges into an expansion tank, as in modern sealed cooling systems. In these systems water is drawn back into the radiator header tank to relieve the partial vacuum that develops as the engine cools down.

A spring-loaded valve incorporated in the radiator filler cap of a normal system, or in the expansion-tank cap of a fully-sealed system opens only when the pressure in the system exceeds 4–7 or 11–13 lb per sq in. A second valve, operating in the reverse direction, prevents the development of a semi-vacuum as the system cools down and water vapour condenses.

A further advantage of the pressurized system is that the boiling point of the coolant is raised by about 1·6°C (3°F) for each pound of pressure increase, thus giving an additional safety margin and preventing local boiling at overheated points

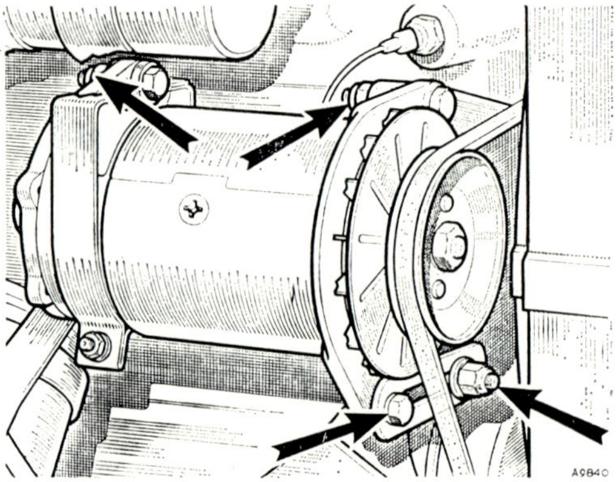

FIG. 62. TO ADJUST THE DYNAMO AND FAN BELT TENSION, SLACKEN THE NUTS INDICATED BY THE ARROWS

in the cylinder head when the engine is operating under very arduous conditions.

When excessive loss of water from the system is experienced, the first point to check is the washer on which the valve portion of the filler cap seats. If renewing this washer does not cure the trouble, the simplest plan is to fit a replacement cap.

Prior to the introduction of the modern form of filler cap, a spring-loaded valve was sometimes fitted in the radiator overflow pipe. If this type of valve should stick in the closed position, the pressure generated in the system may be sufficient to burst a water hose or damage the radiator. Alternatively, should the valve fail to allow air to be drawn into the header tank as the engine cools down, the hoses will collapse. It is, however, usually a simple matter to dismantle the valve.

Apart from leakage from the hose joints, the water pump

gland, cylinder head gasket, or a defective radiator—faults which are discussed later in this chapter—water may also be lost through the overflow pipe owing to surging under normal driving conditions. To prevent this, most modern systems are fitted with some form of baffle or a spring-loaded valve in an overflow pipe which prevents loss of water, but allows air pressure or steam to escape.

Radiator Filler Cap Troubles. When removing the cap from a pressurized system always first turn it as far as the safety notch to allow pressure and steam to escape. If the cap is removed quickly from a very hot engine boiling water may be sprayed out of the filler, with the risk of scalding the hand.

A simple method of testing the cap is to remove it from a hot engine; the release of pressure will then be clearly audible if the cap is sealing the system properly—unless, of course, there is a leak in some other part of the system, for example at a hose connexion.

Examine the cap for worn or damaged parts, paying particular attention to the condition of the locking diaphragm and sealing gasket. There are two types of pressure cap in common use: the "open" type, in which the larger pressure release spring is clearly visible and the rubber sealing gasket is an integral part of the cap, and the "closed" type in which the pressure release spring is shrouded and a separate gasket is used in the filler neck to form the seal. This gasket should always be removed, examined for wear and if necessary replaced when the cap is checked.

If the system is otherwise free from leaks, the test suggested above will indicate whether or not the cap is retaining pressure. On the other hand it cannot reveal loss of spring tension which causes the cap to release pressure at a lower pressure, nor is an engine always conveniently hot enough to test the cap in this way.

For really positive testing, the AC pressure cap and cooling system tester must be used. This is used by the majority of garages.

Testing for Gas or Air Leakage into System. Loss of cooling water may be caused by leakage of exhaust gas or compressed mixture into the system due to a faulty cylinder head gasket, as described under the heading "Signs of Gasket Leakage" on pages 43–4. This will create pressure in the water jackets

CHART No. 10

COOLING SYSTEM DEFECTS

- **Engine Overheats**
  - **At Normal Atmospheric Temperatures** — Check radiator water level — Check tension of fan belt — Check operation of thermostat (see text) — If foregoing points are satisfactory, cause may lie in:
    - Radiator and cooling system faults
      - Choked radiator passages.
      - Radiator partly blanked off.
      - Defective hose connexions.
      - Water pump defective.
      - Loose material in header tank.
      - Fan blades bent.
      - Clogged jets in water distribution tube (see text).
      - High-altitude operation.
    - Engine and chassis faults
      - Ignition timing retarded (Chart No. 6).
      - Weak mixture (Chart No. 8).
      - Bearings or pistons tight (Chart No. 3).
      - Pre-ignition (Chart No. 3).
      - Silencer choked (Chart No. 3).
      - Excessive tractive resistance (Chapter I).
      - Brakes binding (Chapter X).
  - **In Very Cold Weather** — Check radiator water level — If correct and if anti-freeze is not used:
    - Radiator frozen at base.
    - Water pump spindle sheared due to frozen pump.
- **Engine Overcooled**
  - **At Normal Atmospheric Temperatures** — Check operation of thermostat (see text)
    - Thermostat satisfactory
      - Wrong diagnosis—water temperature thermometer faulty.
      - Radiator too large—partly blank off or remove fan blades (see text).
    - Thermostat faulty — Thermostat remains open.
  - **In Very Cold Weather** — Check operation of thermostat (see text) — If satisfactory — Blank off part of radiator or remove fan blades (see text).
- **Loss of Cooling Water** — Examine for visible leaks and tighten all joints — If no visible water leaks, test for gas leakage and air suction (see text)
  - Radiator overfilled (see text).
  - General overheating and boiling.
  - Local boiling in cylinder head causes loss from overflow (see text).
  - Defective radiator: leaks only when subject to road vibration.
  - Cylinder head gasket leaking (Chart No. 3).
  - Faulty overflow relief valve

which will raise the level of the water in the header tank and cause loss of water from the overflow. In addition, water may also leak into the cylinders when the engine is stationary, as also described under the same heading.

As a more positive test than that described on page 44, it is advisable to drain the radiator until the water level is just below the top of the cylinder head offtake pipe or thermostat housing. Remove the upper radiator hose, and the thermostat (if fitted) and take off the fan belt to prevent the water pump operating. If necessary, add sufficient water until the level is just below the open end of the pipe.

Starting with the engine cold, run it at a fast speed and watch for bubbles in the water in the pipe, both at a steady engine speed and also when the engine is suddenly accelerated and allowed to drop back to idling speed. While the test should be made over as wide a temperature range as possible, it should be discontinued before boiling starts, as the formation of steam bubbles will give a misleading impression.

Sometimes air can be drawn into the cooling system by pump suction through a leaking water pump gland, or faulty hose connexions or flange joints. Air in the system will cause rapid corrosion, so that it is as well, when this trouble is experienced, to make a test by removing the overflow relief valve and attaching a length of rubber tube to the base of the overflow pipe.

Top-up the header tank to the normal level, allowing room for expansion, and make sure that the filler cap is air-tight. Run the engine until the temperature no longer rises, and then dip the end of the rubber tube in a glass of water. Continuous bubbles indicate that air is being drawn into the system. If the temperature of the cooling water is still rising, however, the expansion of the water will force air out of the header tank and thus give misleading results. Similarly, if the water is nearing boiling point, steam bubbles will appear. The test is best made, therefore, on cooling systems fitted with a thermometer, so that it can be verified that the temperature is constant.

Checking Thermostat Operation. When overheating is experienced the thermostat should be checked. Most modern engines incorporate this aid to quick warming-up. Whereas early thermostats were often fitted in the upper radiator hose connexion, where their visibility acted as a reminder of their presence, thermostats nowadays are usually incorporated in the water offtake casting or combined with the water pump housing, so that their existence is often forgotten.

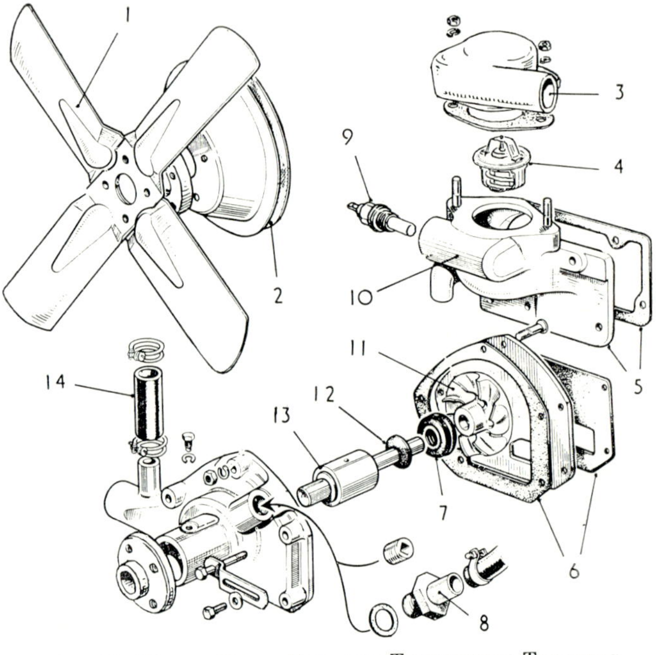

Fig. 63. Water Pump, Fan and Thermostat Troubles

1. Fan blades bent or broken
2. Fan pulley damaged
3. Water outlet pipe leaking
4. Thermostat defective
5. Gasket leaking. Flange distorted
6. Cover plate distorted. Gasket leaking
7. Water seal leaking
8. Heater hose or blanking plug leaking
9. Thermostat element defective
10. Thermostat housing damaged
11. Impeller corroded
12. Thrower damaged
13. Bearing worn or squeaking
14. By-pass hose softened or leaking

Modern thermostats consist of a main valve which controls the flow of cooling water, and sometimes a subsidiary valve which shuts off a by-pass when the main valve is fully open. The by-pass is required, when pump circulation is used, to allow the water to circulate back to the cylinder block without passing through the radiator, thus ensuring rapid warming up without causing an excessive pressure in the system. Both valves are carried on a spindle which is actuated by a wax-filled capsule or by a collapsible metal bellows filled with a volatile liquid that expands the bellows when a critical temperature is reached.

On modern engines this is usually 82°C (180°F) for temperate climates, 88°C (190°F) for winter use in cold climates and 74°C (165°F) for tropical conditions.

The thermostat can cause overheating, therefore, only if the main valve fails to open, or the by-pass remains open, although the latter fault is not usually very serious when an efficient water pump is fitted.

Why Thermostats may Cause Boiling. Nevertheless, a thermostat can cause overheating if the valve spindle becomes so corroded or encrusted with lime deposits that it no longer works freely, or if deposits on the valve and seating seriously restrict the effective area of opening. Another possible trouble is the result of refilling a cooling system too quickly after it has been completely drained. The thermostat valve being closed, and the water pump stationary, an air lock is likely to form, so that the radiator header tank appears full when in fact the cooling system contains only a small quantity of water.

If the car is driven away immediately, the water will quickly boil, and the resulting temperature reached by the cylinder head may be high enough to melt the solder in the thermostat, putting it completely out of action. When the boiling is discovered and the radiator header tank correctly topped up, boiling will of course quickly recur. If, at this stage, the thermostat is examined and found to be useless, it will probably be unjustly blamed for the initial overheating.

Testing the Thermostat. After the removal of a thermostat by unbolting the flange of its housing from the cylinder head or block, or the extraction of the drop-in type of thermostat after removal of the circlip which secures it, the unit should first be visually examined for mechanical damage and excessive deposits of lime. If it appears to be in order, the main valve being closed, it should be placed in water heated to about 11°C (20°F) below the temperature marked on the thermostat flange. The water should then be heated while moving the thermostat about and the valve should begin to open at a temperature within 3°C ($5\frac{1}{2}$°F) of that marked on the flange and should be fully open (at least $\frac{1}{4}$ in. valve lift) within a further rise of 11–14°C (20–25°F).

It must be remembered, however, that thermo-syphon-cooled engines often run at a considerably higher temperature

FIG. 64. REMOVING THERMOSTAT (*left*) AND TESTING IT IN HOT WATER (*right*)

than do pump-cooled engines, frequently exceeding 95°C in hot weather. The thermostat, therefore, should not be blamed for this higher temperature.

Deposits in Water Passages. If the thermostat is proved to be in order, the cause of overheating must lie in the radiator, hoses, water pump or cylinder block water passages. The most usual cause is an accumulation of lime and rust deposits which chokes the small waterways in the radiator. Normally most of these deposits are carried around in suspension in the cooling water, but if a badly furred radiator is allowed to dry out, the deposits harden on contact with air and do not soften again when the radiator is refilled. On many modern engines, too, high-velocity jets of cooling water are directed upon critical points in the cylinder head and block, such as the exhaust valve seatings and sparking plug bosses, by an internal distributor tube or by a series of small openings in the cylinder

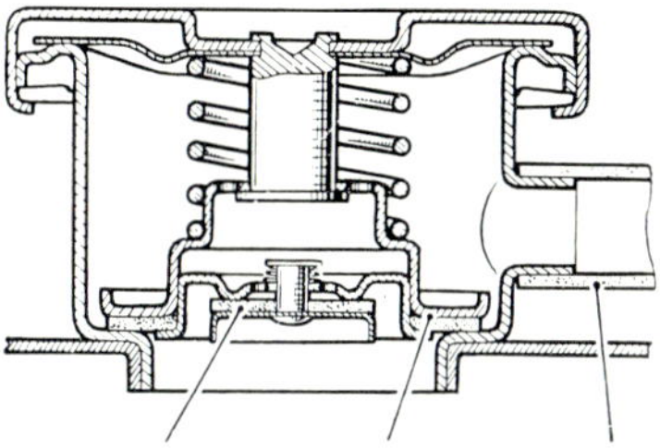

FIG. 65. SECTION THROUGH A RADIATOR PRESSURE CAP

block casting. If these small jets become choked, detonation and pre-ignition can be expected.

The deposits may be removed by the use of a proprietary radiator cleansing solvent. Such solvents are preferable to home-made solutions of soda-ash, caustic soda, or oxalic acid, which are apt to corrode the cooling system components. The solvent having been used as directed by the manufacturer, the system should be thoroughly flushed out, the cylinder block and radiator being separately dealt with, and water forced through the passages in the reverse of the normal direction of flow.

Some car manufacturers recommend the use of a corrosion-inhibitor in the cooling system. Such inhibitors take the form of soluble oils which deposit a protective film over all the internal surfaces of the system, or chemical compounds which are very effective in retarding the formation of deposits.

If an inhibitor is used in conjunction with an anti-freeze mixture, however, some caution is necessary, as the combination of a given inhibitor with a particular brand of anti-freeze may result in the formation of sludge in the system.

Again, if the anti-freeze already contains an inhibitor, the double dose may have a harmful effect on rubber hoses and gaskets. The advice of the car manufacturer should, therefore, be followed.

Boiling and Overheating in Cold Weather. A further point to bear in mind is that anti-freeze solutions have a very searching action in loosening any rust and scale in the cooling system; there are generally heavy deposits in the cylinder

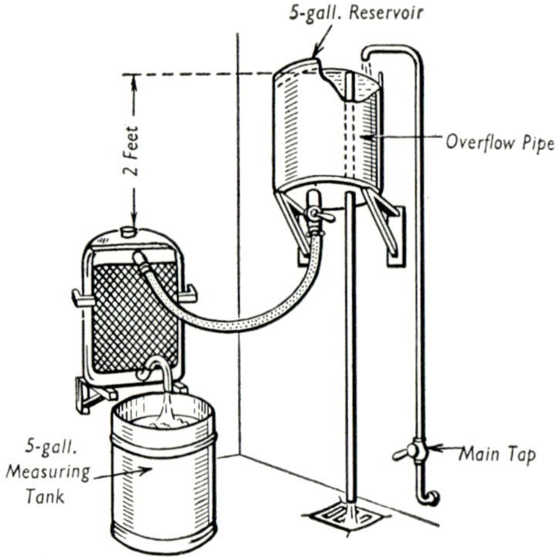

FIG. 66. FLOW TEST APPLIED TO VAUXHALL RADIATORS AFTER REPAIR

block, where temperatures are highest. If anti-freeze is used without first thoroughly cleaning out the system with a solvent and reverse-flushing the cylinder block and radiator, the deposits loosened in the cylinder block will be carried to the header tank, where they may choke most of the radiator tubes. To blame the anti-freeze solution for this state of affairs is faulty diagnosis: if the cooling system is clean before the anti-freeze is added, the solution will keep it clean.

The searching action of the solution will similarly dissolve any deposits which may be sealing defective water joints, leading to the appearance of leaks. The remedy is, of course, to tighten all hose joints, flange nuts, and pull down the cylinder head nuts evenly in the sequence specified by the manufacturer while the engine is thoroughly hot.

If anti-freeze is not used, there is always the risk of the water in the system freezing in frosty weather, probably resulting in a cracked cylinder block or a burst radiator core. Although the car may be kept in a heated garage, or an under-bonnet or under-sump heater may be used to prevent freezing, there still remains the risk of the water at the base of the radiator freezing during the first few miles of driving on the road. Since the thermostat restricts the water circulation to the cylinder block until the engine is thoroughly warmed up, the fact that the radiator freezes is not surprising. This results in violent boiling once the engine is warm, as the ice prevents circulation.

Another cause of boiling in winter is a sheared water pump spindle. Even though the radiator may be drained overnight, water may be trapped in the pump casing ; a drain tap is normally fitted to the pump if there is any risk of this occurring, and if it is overlooked the ice which forms will effectively lock the impellor. When the starter is used it may be unable to turn the engine ; alternatively the pump drive spindle may shear, especially if the car is towed to start it.

Rubber Hose Troubles. While a perished rubber hose will be obvious on external inspection, internal defects in an apparently sound hose can often be overlooked. A possible cause of overheating, for instance, is a partially detached section of the lining which acts as a flap valve and restricts the flow of water through the hose. Sections of lining may also become detached from the upper hose and pass into the header tank, blanking off a number of water tubes, or may enter the water pump or cylinder jackets from the lower hose.

Another possible trouble, which is evident only when the engine is running at high speed, is a perished lower hose connexion which collapses under the suction exerted by the water pump.

Water Pump Faults. A faulty water pump can be rectified only by dismantling it and fitting replacement parts ; consequently the manufacturer's instructions must be obtained. Leakage from the pump gland can be rectified at the same time, since most modern pumps have self-adjusting packings or carbon gland rings which must be renewed when leakage takes place. If the gland is adjustable, however, slight tightening of the adjusting sleeve or gland nut may be all that is required to cure the leak. Overtightening the gland will cause rapid wear of the packing and may score the pump spindle.

Testing a Radiator. Radiator faults usually call for the attention of a specialist. Inexperienced attempts to repair a leak with a soldering iron often result in adjacent soldered joints being melted and the area of damage increased. After repair it is usual to test the radiator under air pressure while it is immersed in a tank of water. Generally this test is followed by a flow test, during which the time taken for a given quantity of water to flow through the radiator, from a tank at a specified height above it, is accurately measured. As a rough guide, it can be taken that, a four-foot head of water being used, five gallons should pass through the average radiator in 20–30 seconds, but whenever possible the exact figures for time and quantity, and the correct head of water, should be ascertained from the manufacturer.

CHAPTER VII

DIAGNOSING TRANSMISSION TROUBLES

THE testing and servicing of a modern transmission system, which may include an automatic clutch or gearbox, is a very wide subject deserving a volume to itself, for example, *Automobile Transmission Servicing and Overhaul*, a companion volume in this series which covers in detail practical aspects of the work which can be referred to only briefly in this chapter. Correct diagnosis of the causes of noise, vibration, backlash or faulty operation, moreover, usually calls for a considerable amount of experience. When the parts of, say, a gearbox or the final-drive and differential assembly have been dismantled and are laid out on the bench, it may be difficult to determine which items are sufficiently worn to cause the troubles experienced in service—excepting, of course, such obvious faults as broken or chipped gears, pitted or cracked ball races, broken balls and so on.

The position is complicated, in the case of a two-pedal control system or a fully-automatic transmission, by the relative complexity of the control system. It will normally be essential to have available for reference a diagram of the layout of the system, together with a schedule of the correct adjustments and tests. In the case of an automatic transmission, a list of the road speeds at which upward and downward gear changes should occur under various driving conditions will be essential if fault-tracing is to be carried out systematically. It must be emphasized that haphazard tests, and tinkering with any adjustments that may be accessible, are likely to lead only to further trouble.

Therefore in considering the extent to which it would be advisable for the average owner or mechanic to attempt to carry out any extensive work on the components of the transmission system, we are forced to the conclusion that in most cases he should confine himself to isolating the trouble in the clutch, gearbox or automatic transmission, propeller shaft or final-drive components, and then to narrow, as far as possible, the choice of faulty components within that unit.

This is a logical approach, since the growing tendency today, when dealing with units which call for the use of special service tools and fixtures in dismantling and reassembly, is to take full advantage of the service-exchange scheme operated by most major car and component manufacturers, whereby a faulty component or assembly can be exchanged for a fully-reconditioned and guaranteed replacement.

In practice, the scheme has a great deal to recommend it; usually it proves both cheaper and more satisfactory in the long run. Sometimes, one has no option in the matter: on automatic transmissions, for example, a complex item such as the torque converter is often of welded construction and cannot be dismantled for inspection or repair.

Transaxle Designs. The growing trend to combine the engine, gearbox and final drive gears in a single unit on front-wheel-drive or rear-wheel-drive cars further complicates fault-tracing and servicing. Although the units themselves—the clutch, gearbox, differential and reduction gears—are often of the same type as those used when the various units of the transmission system are separate, it is obviously more difficult to diagnose, for example, the source of noisy operation when the gearbox and final drive gears are combined in one casing.

Two-pedal Controls. This term is usually applied to semi-automatic transmissions in which the clutch is automatically engaged and disengaged whenever the gear-lever knob is grasped, and in which some measure of control over the engine speed is provided in order to synchronize the gear changes, thus enabling the car to be controlled by the accelerator and brake pedal alone.

These controls can be divided into two broad classes: those in which a single-plate friction clutch is automatically engaged by centrifugal weights whenever the engine speed rises above the normal idling setting, and is disengaged by some form of vacuum-servo when the gear-lever knob is grasped, thus allowing gear changes to be made when the engine is running at normal speeds and the centrifugal mechanism would otherwise keep the clutch engaged, examples being the Manumatic and Newtondrive systems; and electro-magnetic clutches, such as the Ferlec, in which magnet coils embedded in the flywheel attract an armature which applies pressure to a friction-plate assembly, or the Smiths or Jaeger design in which the drive is conveyed from the flywheel to the driven member of the clutch through magnetic particles which are trapped in a small air-space between the two members.

In electro-magnetic clutches advantage is taken of the rising output of current from the engine dynamo, as the engine speed is increased from the idling setting, to engage the clutch. Disengagement is brought about when necessary by a switch incorporated in the gear-lever knob, controlling a relay which interrupts the current supply to the clutch magnet windings. Further refinements are incorporated in the control system: for example, resistances may be brought into circuit, depending on the gear ratio in use and the throttle opening, to ensure smooth engagement of the clutch under a wide variety of operating conditions and there is usually provision for operation of the clutch directly from the battery should the dynamo fail.

Fault-tracing On Two-pedal Control Systems. From what has just been said, it will be evident that it is impossible to give any general instructions that would be applicable to all types of clutch and control systems. Fortunately, the most widely used two-pedal controls are based on proprietary units. The Newtondrive system was fitted to the smaller Ford and Standard cars, for example, whereas the Manumatic control was optional on some B.M.C. models—chiefly Austin and Morris—and also on Hillman Minx, Armstrong Siddeley and some other cars. It should be possible, therefore, to obtain details from the local agents for the car in question or from the manufacturer's service department.

Broadly speaking, the mechanically-operated automatic clutches which form part of the systems just mentioned are basically similar to conventional single-plate clutches which are dealt with in Chart No. 11; it is impossible to consider the majority of the faults listed in the chart under "Faulty Engagement," however, without taking into consideration the control system of a two-pedal installation. The design of the clutches, too, is a little more complex than the conventional type.

With electro-magnetic clutches, fault-tracing is often confined to carrying out a systematic series of electrical tests at various points in the control circuit. Mechanical faults cannot be entirely ruled out, however, especially in the case of the

Ferlec clutch, on which faulty engagement may be attributable to wear on the friction plate, pressure plate or intermediate plate, or on the linkage coupling the parts to one another and to the flywheel.

Automatic Transmissions. The majority of transmissions are based on American designs in which a fluid flywheel or a torque converter couples the engine to an epicyclic gearbox in which the appropriate gear trains are brought into action by brake bands or multiple disc clutches operated by oil pressure.

A complex control system, linked to the accelerator pedal and incorporating a governor which is driven from the output shaft of the gearbox, contains a series of valves that direct oil pressure to the appropriate brake bands and clutches, thus bringing a suitable gear ratio into use, as determined by the amount of throttle opening, the engine and road speed and the load on the engine.

Thus upward changes are made at relatively low road speeds if only a small throttle opening is used, but are delayed until the engine reaches peak revolutions in each gear if the throttle is held wide open during acceleration. Usually the driver is given an overriding control which allows him to select or retain a lower gear ratio than that which would be selected by the automatic transmission under certain running conditions, as when descending a hill with the throttle closed, and also a kick-down control that allows a lower gear to be instantly engaged by pressing the throttle pedal beyond the normal limit of its travel.

A typical test schedule will usually include a "creep" test, to reveal any tendency of the car to creep forward when the transmission selector lever is moved to any of the driving ranges, followed by a "stall" test, which takes the form of running the engine at full throttle with the selector in a driving range and the wheels braked. This test, which calls for the use of an engine-revolution indicator of the type fitted to most modern test-tune sets, will reveal whether the engine power is below standard, as indicated by the failure of the engine to reach the specified speed, or whether any slip is taking place in the fluid coupling, torque converter or brake bands of the transmission, shown by the engine exceeding the specified stall speed.

A very comprehensive road test is then carried out, in which

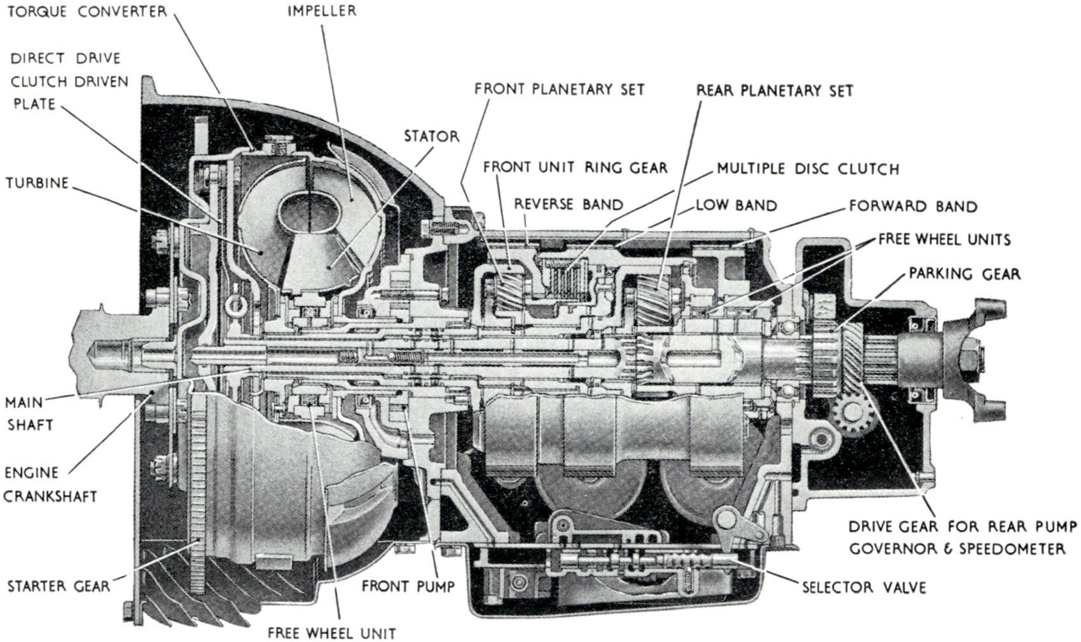

FIG. 67. THE COMPONENTS OF THE BORG-WARNER AUTOMATIC TRANSMISSION

the behaviour of the transmission under all driving conditions is logged on a report sheet and subsequently checked against the manufacturer's specification.

If faults are revealed by any of these tests, the next stage is to carry out a series of oil-pressure tests, an oil-pressure gauge, provided with a long flexible pipeline, being connected to various pressure take-off points on the transmission which are normally sealed by screwed plugs. Tests are made at different speeds and under different driving conditions, the results of which will help the expert to determine whether faults are due to derangements of the hydraulic system—for example in the valves, governor or servo pistons and cylinders—or whether they are caused by mechanical defects, such as brake bands which have worn to such an extent that adjustment or replacement is required.

B.M.C. Mini and 1100 Models. The automatic transmission which is an optional alternative to the manually-controlled gearbox on B.M.C. Mini and 1100 models is unusual in that two sets of bevel reduction gears are used to provide four ratios, any one of which can be selected and held by the driver; alternatively, with the selector lever at "D," fully automatic gear changing, in response to engine and car speed and accelerator pedal position, is obtained.

The use of bevel reduction gears results in a very compact assembly, enclosed within the transmission case of the engine, together with the differential and final drive gears. A torque converter replaces the flywheel and clutch assembly, and power is transmitted through an idler gear to the input gear of the gear cluster. The general arrangement is shown in Fig. 68.

Servicing and fault-tracing procedures are described in the appropriate workshop manual. Insofar as regular maintenance is concerned, the chief point to bear in mind is that the engine oil is used also in the torque converter and in the hydraulic control system; as a result it is essential to change the oil regularly at 3,000-mile intervals, or more frequently if the engine is operating under adverse conditions such as frequent starts from cold and short runs. When the oil begins to deteriorate, the friction characteristics of the clutches and brake bands are altered, interfering with the smooth operation of the transmission.

It is also important to change the oil filter element at the

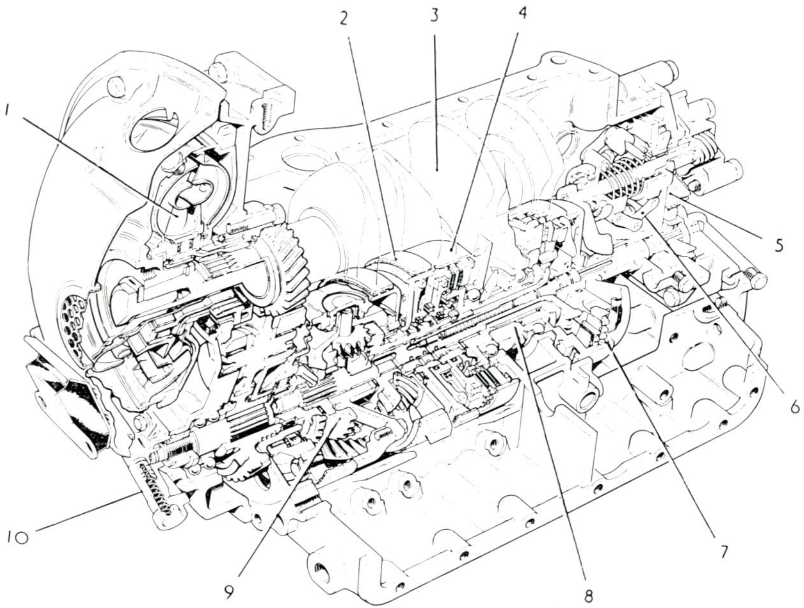

Fig. 68. Automatic Transmission used on B.M.C. Mini and 1100 Ranges

The torque converter replaces the normal clutch and the gears are housed in the transmission case beneath the engine crankshaft

1. Torque converter
2. Second and third gear brake bands
3. Engine crankshaft
4. Top and reverse gear clutch
5. Auxiliary oil pump
6. Governor
7. Forward clutch
8. Final-drive pinion
9. Bevel epicyclic gears
10. Low-pressure valve

specified mileage (normally 6,000 miles) since contaminated oil will seriously affect the operation of the transmission.

Smith's Automatic Transmission. This transmission—known in Europe as the Jaeger design—is particularly suitable for smaller cars powered by engines of about $1\frac{1}{2}$-litres capacity, operates on a different principle from those just

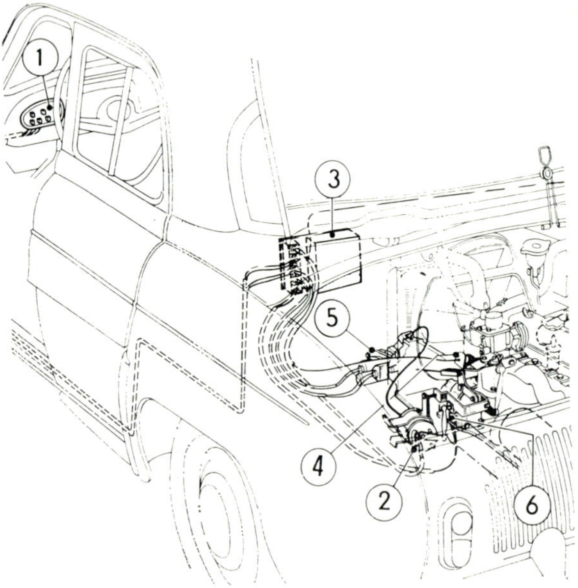

FIG. 69. CONTROL SYSTEM OF TYPICAL SMITHS-JAEGER TRANSMISSION

1. Press-button selectors
2. Governor responding to engine speed
3. Relay box
4. Secondary throttle
5. Gear selector actuator
6. Magnetic-powder coupling

described. As will be seen from Fig. 70, the transmission consists of two magnetic-powder couplings which drive a three-speed and reverse synchromesh gearbox.

As indicated earlier in this chapter, a magnetic-powder coupling can be used as an automatic clutch. In this case, the rear clutch drives the constant-mesh gears of the gearbox in the conventional manner, thus rotating the layshaft and providing a smooth take-off from a standstill and operation in first and second speed ratios, depending on which gear is locked to the mainshaft of the gearbox by the usual synchromesh and dog-clutch mechanism. The latter is operated by the gear-engagement solenoid, which is energized at the appropriate moment by the control system. Top gear is a direct drive from the forward magnetic clutch, through an extension of the gear-box mainshaft which passes through the hollow primary shaft; the drive is thus obtained by switching off the rear magnetic coupling and energizing the forward coupling. Again, of course, this is done automatically by the control system.

Fault-tracing with this type of transmission is a combination of electrical tests of the control system, the magnetic-powder clutches and the gear-engagement solenoid, and mechanical checks on the synchromesh gearbox assembly, which closely resembles the conventional type of gearbox dealt with in Chart No. 12.

Overdrives. During recent years, the overdrive, in the form of a "step-up" gear interposed between the gearbox and the final-drive gears, has become increasingly popular, either as a standard fitting or as an optional extra. The designs most widely used on British cars—the Laycock de Normanville and the Borg-Warner—consist of planetary gears which are brought into action by a solenoid. On many Laycock de Normanville installations, the driver can engage or disengage the overdrive at will by operating a switch mounted on the steering column or on the instrument panel. In some cases, however, the switch merely preselects the overdrive, actual engagement and disengagement being brought by a control unit which ensures, by means of a governor switch, that the overdrive will come into action only when the speed of the car exceeds about 30 m.p.h. in top gear and will disengage automatically below this speed, thus avoiding any risk of the engine being forced to labour in too high a gear at low speeds. In addition, a throttle-operated switch may be included to prevent overdrive being engaged when the throttle is closed, as this type of overdrive engages most smoothly when the engine is pulling.

With either the manual or the automatic control system, a master switch, operated by the gear-lever linkage or by the selector mechanism in the gearbox, ensures that overdrive will operate only when the appropriate gears have been selected:

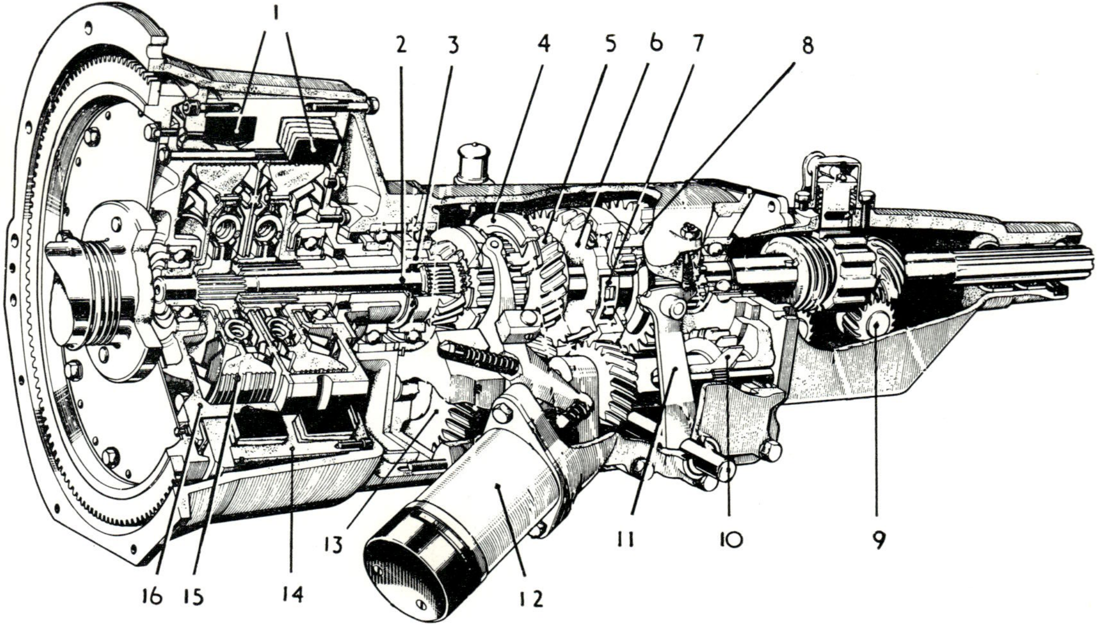

FIG. 70. SMITHS "EASIDRIVE" TRANSMISSION, AS FITTED TO THE HILLMAN MINX AND SINGER GAZELLE

1. Magnet coils
2. Gearbox mainshaft
3. Gearbox outer shaft
4. Dog clutch
5. Intermediate gear
6. Low gear
7. Free-wheel
8. Reverse gear
9. Governor drive
10. Reverse idler gear
11. Reverse operating lever
12. Gear-engagement solenoid
13. Layshaft gear
14. Stator
15. Driven member
16. Driving member

usually top and third gear of a four-speed gearbox, or top and intermediate gear of a three-speed box. The power transmitted through a lower ratio in each case may overload the overdrive gears; moreover the overdrive will be seriously damaged if the car is driven in reverse with the overdrive engaged. Correct adjustment of the gear selector switch is, therefore, of vital importance.

The Borg-Warner overdrive is usually operated automatically, coming into action at road speeds above 30 m.p.h. in top gear and disengaging when the speed falls below about 27 m.p.h. A kick-down switch, operated by depressing the accelerator pedal beyond its normal range of travel, enables overdrive to be cut out instantaneously when maximum acceleration is required. In most installations, overdrive is available on at least one intermediate gear, in addition to the top gear ratio.

Tracing Overdrive Faults. The overdrive unit itself seldom gives trouble and has a very long life between overhauls. Such faults as do crop up can usually be traced to the control system, the operating solenoid and the external wiring. First, however, check that the gearbox is topped-up to the correct level with the recommended grade of oil. Shortage of oil, or an incorrect grade, can cause trouble, since the overdrive draws its supply from the gearbox. On the Laycock de Normanville unit, the clutch that engages the overdrive is hydraulically operated and if the oil is too thick, the action may be sluggish. Even with the correct oil in use it is not uncommon, in very cold weather, for the overdrive to be reluctant to engage during the first mile or so after starting from cold.

In carrying out electrical checks the first requirement is, of course, a wiring diagram for the particular installation, although this may not be essential in the case of a Laycock single-relay system with direct manual control by the driver.

With a voltmeter or test lamp, check that current is reaching and passing through each switch, relay or solenoid in turn, as indicated below. A suspect item can be temporarily short-circuited with a jumper wire or tested by substitution if a serviceable spare is available. Never be tempted to interfere with any of the pre-set adjustments on the overdrive: such work should be done only when the appropriate service manual is available.

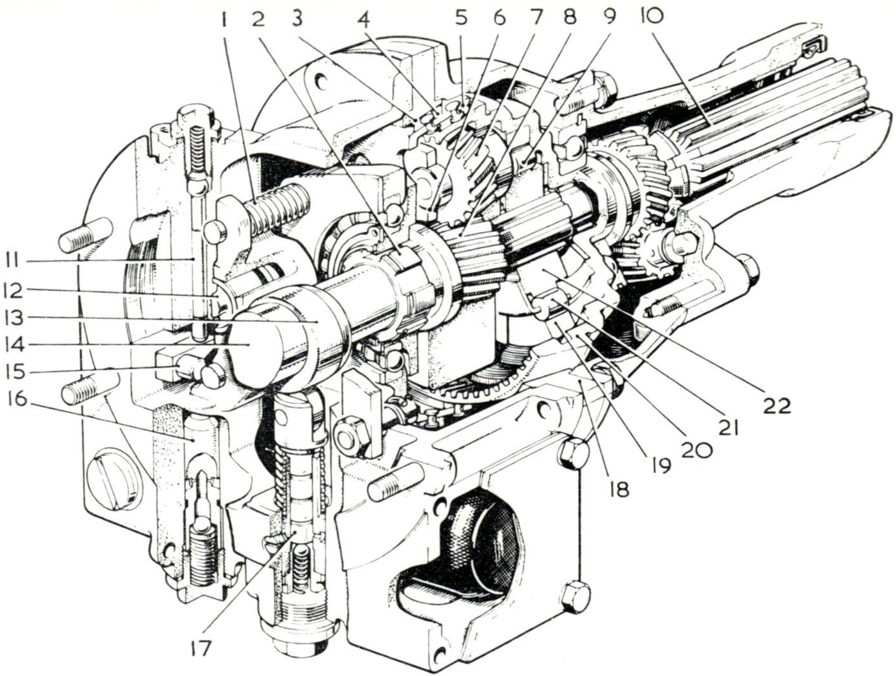

FIG. 71. TYPICAL LAYCOCK DE NORMANVILLE OVERDRIVE

1. Cone clutch spring
2. Splined sleeve
3. Outer friction lining
4. Cone clutch
5. Inner friction lining
6. Planet carrier
7. Planet wheel
8. Sun wheel
9. Free-wheel
10. Output shaft
11. Operating valve
12. Operating piston
13. Cam
14. Input shaft
15. Operating shaft
16. Non-return valve
17. Plunger pump
18. Brake ring
19. Outer member of free-wheel
20. Annulus
21. Free-wheel roller
22. Inner member of free-wheel

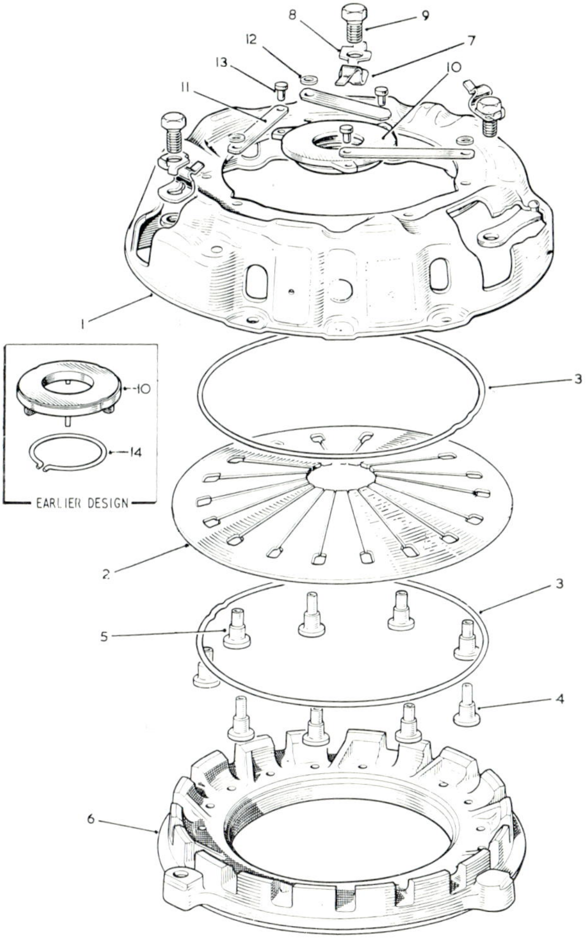

FIG. 72. TYPICAL BORG AND BECK DIAPHRAGM-SPRING CLUTCH

The following are some faults to look for—

1. Cover distorted
2. Diaphragm spring broken
3. Fulcrum rings broken or distorted
4. Locating pins worn
5. Shouldered rivet loose
6. Pressure plate worn or scored
7. Retractor clip broken
8. Tab washer not locked
9. Bolt loose
10. Clutch release plate worn
11. Driving strap broken
12. Washer missing
13. Rivet loose
14. Retaining ring broken or distorted

CHART No. 11

CLUTCH TROUBLES

- FAULTY ENGAGEMENT
  - SLIP
    - Check free movement of pedal and adjust if necessary
      - If correct, dismantle clutch to check for following faults:
        - Oil or grease on friction linings.
        - Release mechanism binding.
        - Weak thrust springs: heat caused by excessive slipping will soften springs and aggravate trouble.
        - Release levers binding on eyebolt pins.
  - FIERCENESS OR SNATCH
    - Accurate diagnosis calls for dismantling clutch
      - Oil or grease on friction linings.
      - Friction linings worn out.
      - Broken or misaligned pressure plate.
      - Hub binding on splines.
      - Governor weight levers binding in automatic clutch (see text).
  - JUDDER
    - Accurate diagnosis possible only on dismantling clutch
      - Linings not making even contact.
      - Pressure plate not parallel with flywheel.
      - Buckled driven plate.
      - Bent splined shaft.
  - DRAGGING OR SPINNING
    - Check clutch adjustment check grade of oil in oil-filled clutch
      - If correct, dismantle clutch: trouble due to one or more of following faults:
        - Oil or grease on friction linings.
        - Clutch shaft out of line with engine.
        - Pressure plate warped or damaged.
        - Hub binding on splines.
        - Driven plate distorted.

Symptom	Type	Test / Condition	Result	Cause
Noisy Operation	Rattle	Depress clutch pedal to take up free movement, but not sufficiently to compress springs	Rattle disappears	Worn or loose graphite release block. Worn release bearing parts. Weak or broken anti-rattle springs on toggle arms. Pedal return spring disconnected.
			Rattle continues	Damaged driven plate. Bent splined shaft. Excessive backlash in transmission.
	Tinkling or High-Pitched Whine	Tinkling noise is evident when clutch is engaged with engine idling; disappears when pedal is depressed. May be replaced by high-pitched whine.		Ball thrust bearing worn. Unlubricated ball thrust bearing.
	Knock or Tick	Evident when clutch is engaged with engine idling		Hub splines badly worn—generally due to misalinement. Spigot bearing worn.

FLUID FLYWHEELS

Symptom	Cause
Excessive Slip	Shortage of oil in flywheel. Wrong grade of oil.
Drag when Idling	Wrong grade of oil.
Oil Leakage	Worn gland. Worn bearing.
Vibration	Oil filler plugs changed over, upsetting balance of flywheel. Flange bolts fitted to incorrect holes or nuts changed over on bolts, upsetting balance.

Laycock Overdrives. With a simple manual control system, check the gear-lever switch for continuity and correct adjustment; the control switch; relay; and finally the solenoid on the overdrive casing. On some early overdrives, occasional failure of the solenoid was experienced, due to the entry of

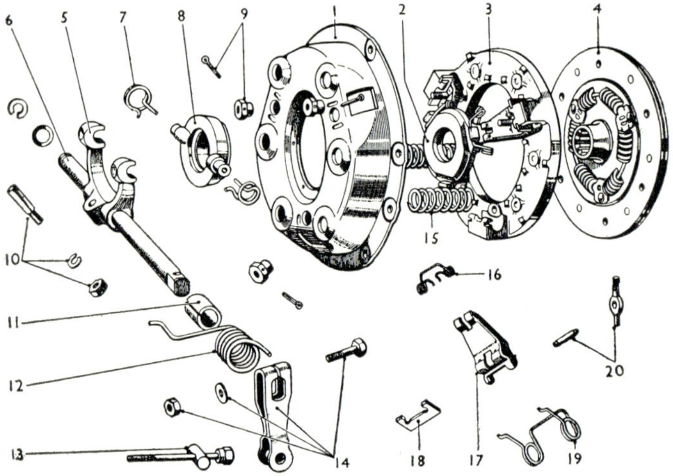

FIG. 73. SOME COMMON COIL-SPRING CLUTCH TROUBLES, MOST OF WHICH ARE COVERED IN THE CHARTS AND TEXT

1. Distorted cover flange
2. Worn or scored release lever plate
3. Scored pressure plate
4. Worn friction linings; distorted plate; weak or broken springs; worn splines in hub
5. Worn or bent withdrawal fork
6. Worn or bent clutch operating shaft
7. Weak or fractured release bearing retaining springs
8. Worn graphite release block
9. Incorrect adjustment of eye bolt nuts
10. Loose cotter pin in withdrawal fork
11. Worn clutch shaft bushes
12. Weak or fractured pedal return spring
13. Worn lever trunnion
14. Loose clamping bolt in clutch shaft lever
15. Weak or fractured clutch pressure springs
16. Defective retaining springs
17. Incorrectly adjusted release levers
18. Displaced struts for release levers
19. Broken anti-rattle spring
20. Worn eye bolts and release lever pins

moisture; sometimes satisfactory operation was restored when the unit had been allowed to dry out.

On two-relay systems, it will be necessary to check, in addition to the items just mentioned, the switch that selects manual or automatic operation (when fitted); the centrifugal switch; the throttle switch, including correct adjustment or the linkage; and the second relay, which is connected to the throttle switch, the centrifugal switch and the solenoid.

Borg-Warner Overdrive. Check the fuse in the relay and the relay itself; the kick-down switch—this has two pairs of contacts, one pair being bridged when the switch is up and the other when it is fully depressed; the adjustment of the switch itself, in relation to the accelerator pedal; the centrifugal governor switch; the solenoid; and, finally, check the adjustment of the control cable between the lock-up lever on the overdrive and the push-pull control in driving compartment.

General Notes on Fault-tracing. Most of the work so far described is, of course, the prerogative of an experienced mechanic, who will probably have attended a special training course. It is likely to be only of academic interest to the average reader, but it does emphasize the point already made: that it is courting trouble to trespass in the province of the expert.

What, then, can the practical owner or mechanic tackle without the benefit of specialized knowledge or service tools? In Charts Nos. 11, 12, 13, 14 and 15, emphasis is placed on first isolating a trouble in the clutch, gearbox, propeller shaft, or rear axle, and then narrowing down, as far as possible, the choice of faulty components in that unit. Conventional rear-wheel drive is assumed, but the basic principles will also apply to front-wheel drive systems, in which the transmission units are similar in spite of their different arrangement.

Transmission Vibration. Vibration, which is often accompanied by noise, is sometimes difficult to track down and cure. The difficulty can be aggravated by the fact that an otherwise satisfactory component is often induced to vibrate in sympathy with an entirely diffierent component; moreover, although the initial vibration may be relatively harmless, the induced vibration may build up to a much more serious extent, particularly at critical speeds.

Assuming, then, that the engine is free from vibration periods caused by uneven compression, carburation, or ignition troubles, or unbalanced components, the quickest method of isolating the trouble in the gearbox, propeller shaft, or rear axle is to drive the car at the road speed at which the vibration is apparent or at its greatest, and to put the gear into neutral.

CHART No. 12

DIAGNOSING PLAIN AND SYNCHROMESH GEARBOX FAULTS

- **Noisy Operation**
  - Test with car stationary and when driving at different speeds in each gear
    - Noise when stationary
      - Worn constant mesh gears.
      - Worn first motion shaft bearings.
    - Noise when driving on one or all ratios
      - Worn gears
      - Worn bearings
  - Second step in diagnosis: note type of noise
    - Whine or hum
      - Worn races.
      - Incorrectly meshed gears due to wrong assembly or thrust washer wear.
      - Worn gears.
    - Knocking
      - Chipped gear tooth.
      - Chipped or broken ball in race.
      - Foreign matter in box.
    - Roughness
      - Worn or pitted races.
- **Stiff Action of Gear Lever**
  - Binding in linkage of steering column gear lever.
  - Gear lever ball joint binding.
  - Faulty synchromesh mechanism.
  - Gearbox cover misalined.
  - Remote control gear lever, linkage misalined.
  - Servo device defective when power assistance provided.

Symptom		Check		Possible causes
SELECTION FAULTS	FAILURE TO ENGAGE GEAR OR NOISY ENGAGEMENT	Determine whether trouble is confined to one ratio or is present on all		Faulty synchromesh mechanism. Selector mechanism faulty. Clutch dragging or spinning (Chart No. 11).
	ENGAGEMENT OF TWO GEARS AT ONCE			Faulty selector mechanism.
	GEARS SLIP OUT OF MESH			Weak selector springs. Worn selector rods, balls or plungers. Worn gears.
	GEARS JAM IN MESH	Fault generally confined to synchromesh boxes: try effect of adding activating agent to oil, such as Redex		Synchromesh cones jammed. Selector fork or rod jammed.
NO POWER TRANSMITTED		Check that trouble does not lie in clutch propeller shaft or rear axle (Charts Nos. 11 and 14)	If trouble is localized in gearbox:	First or third motion shaft fractured. Engagement dog stripped. Constant mesh gears stripped—affect all ratios except top gear. Gears of intermediate ratio affected are stripped.
EXCESSIVE BACKLASH		Check clutch for wear of hub splines (Chart No. 11) check propeller shaft, back axle and hubs for wear (Charts Nos. 14 and 15)	If trouble is localized in gearbox:	Worn gear teeth. Worn selector dogs. Excessive end-play on gears.
OIL LEAKAGE		Check level and grade of oil in gearbox	If correct:	Choked breather. Worn oil seals. Worn bearings.

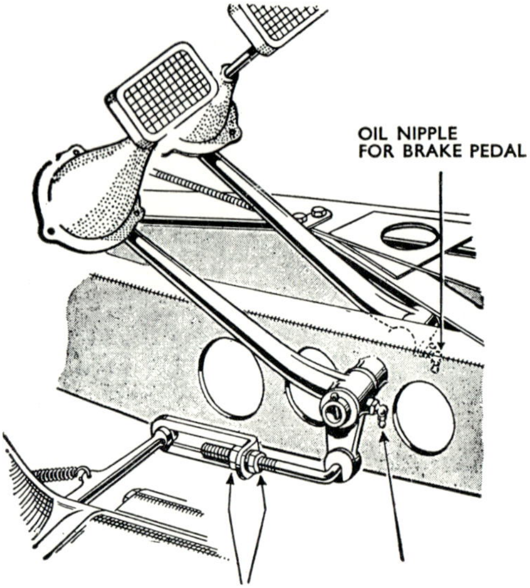

Fig. 74. Widely-used Method of Adjusting the Clutch

When a chain linkage is used instead of the threaded rod shown, judder may occur if the forward end of the chain is not approximately ½ in. higher than the rear. The clutch lubrication nipple should not be overlooked

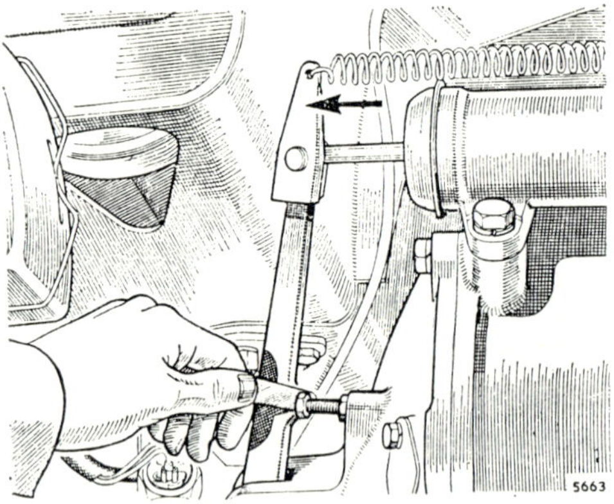

Fig. 75. Measuring Free Movement of Clutch Operating Lever on B.M.C. Mini

Adjustment is by means of the threaded stop

If the vibration disappears, the engine and clutch are, of course, the most likely offenders. If it continues, the propeller shaft and rear axle must be suspected. In practice, the propeller shaft and universal joints are almost invariably to

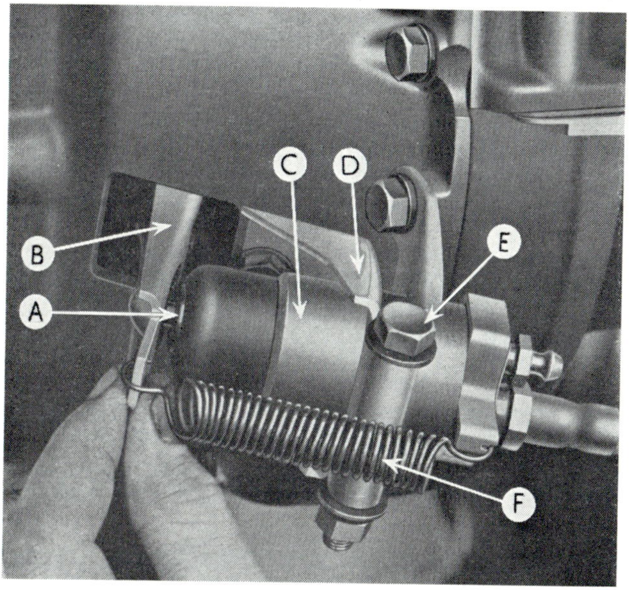

FIG. 76. A COMMON METHOD OF ADJUSTING THE SLAVE CYLINDER OF A HYDRAULICALLY-OPERATED CLUTCH

A. Clearance between push-rod and release lever
B. Clutch release lever
C. Slave cylinder
D. Mounting bracket
E. Clamping bolt
F. Retracting spring

blame, the most usual cause being worn mechanical joints or distorted, torn or softened rubber-bushed joints, especially when these are used at the inner ends of the shafts driving independently sprung wheels. On the B.M.C. Mini, 1100 and 1800 range, for example, oil leakage from the engine can contaminate the rubber segments of the inner universal joints and will greatly shorten their life. The tightness of the U-bolts on these joints should also be checked from time to time since loose bolts will result in rapid disintegration of the joints.

CHART No. 13

PRESELECTOR GEARBOX TROUBLES

BRAKE BANDS SLIPPING
- First check gear change pedal adjustment (see text)
  - Bands slipping on all gears
    - Incorrect pedal adjustment.
    - Pedal binding on shaft.
    - Main spring adjustment incorrect—should need readjustment only in exceptional cases.
  - Bands slipping on one gear
    - "Pump" gear change pedal to take up adjustment
      - If no improvement, check action of automatic adjuster (see text)
        - Adjuster operating
          - Brake band lining worn out.
          - Insufficient toggle leverage, readjustment needed only in exceptional circumstances.
        - Adjuster not operating
          - Adjuster spring weak or broken.
          - Adjusting nut has reached end of thread on pull rod-counter-bore nut.
  - Bands slipping on new or recently reconditioned gearbox
    - "Pump" gear change pedal with appropriate gear selected
      - Friction linings bedding down; automatic adjuster may not compensate for the initial wear.

PEDAL RETURNS WITHOUT GEAR ENGAGEMENT
- Check that gear selector lever is located in notch on quadrant
  - If correctly positioned:
    - Faulty selector mechanism (see below).

FAULTY SELECTION OF GEARS
- Check that gear selector lever is located in appropriate notch on quadrant
  - If correctly positioned
    - Selector linkage incorrectly adjusted.

FIERCE ENGAGEMENT OF GEARS
- Screw out adjuster set-screw of brake band on affected gear one turn.
- Use heavier grade of oil.

TOP GEAR FAILS TO RELEASE
- Slip top gear clutch for short period by engaging gear with handbrake applied while engine is running at 1,500 r.p.m.; allow gear change pedal to come back gently. Repeat at intervals; do not overheat gearbox
  - If trouble is not cured
    - Dirt or foreign matter in oil in gearbox—drain and refill.
    - Clutch cone or plates distorted or damaged.

NOISE WHEN IDLING
- Slight noise is normal when fluid flywheel or centrifugal clutch is not fitted.
  - If gearbox is driven by fluid flywheel or centrifugal clutch:
    - Engine idling speed too high.
    - Clutch or fluid flywheel dragging (Chart No. 11).

The outer constant-velocity joints on cars which have independently-sprung driving wheels are often of the Birfield-Rzeppa type. The chief fault to look for in this type of joint is wear, which may sometimes develop within 20,000 miles of use (or less), causing a knocking sound when the car is driven with the steering turned towards either lock. Slight knocking does not necessarily indicate serious trouble but a loud knock is a warning that there is a risk of the joint failing unexpectedly. New assemblies should, therefore, be fitted before the knock becomes pronounced.

These joints are sealed by rubber gaiters which retain the special grease specified by the manufacturer and which also exclude water and grit. The gaiters should be examined at regular intervals—every 3,000 or 6,000 miles is often recommended—and should be renewed at the first sign of damage. As there is a risk that water or grit may have entered the joint, this should also be carefully inspected for any sign of deterioration before the new gaiter is installed.

A final point to bear in mind when diagnosing faults in constant-velocity universal joints is that although these form part of the steering assembly on front-wheel drive models, failure of the joint itself is unlikely to cause any serious steering defect. When steering troubles occur on cars fitted with this type of transmission, they are more likely to be due to other causes which are discussed in Chapter IX.

Fabric Universal Joint Trouble. Although fabric joints are not common nowadays, two points are worth bearing in mind. In the first place vibration may be apparent only when the engine is pulling, owing to the weakened fabric disks being twisted and distorted when heavily loaded. When this is so, the vibration will disappear when the gear is put into neutral, giving a misleading impression. It is advisable to watch the behaviour of the joints with the car at a standstill, while an assistant slowly engages the clutch with the engine running at a moderate speed with a gear engaged and the handbrake firmly applied. Any distortion of the disks under load will then be evident.

The second point concerns the centering plate which is often fitted to the gearbox and back axle driving yokes of the universal joints so that the opening at the centre of the plate fits over a ball-shaped projection on the end of the propeller shaft and steadies it. If the plate is wrongly assembled by bolting it to the yoke of the joint which is attached to the shaft, on steadying action will be obtained.

Roller Bearing and Bushed Joints. On needle-roller or bronze-bushed universal joints the most usual trouble is wear of the bearings. If the shaft has been removed at any time, however, the splined telescopic portion may be incorrectly assembled on the splines. The yokes on each end of the shaft

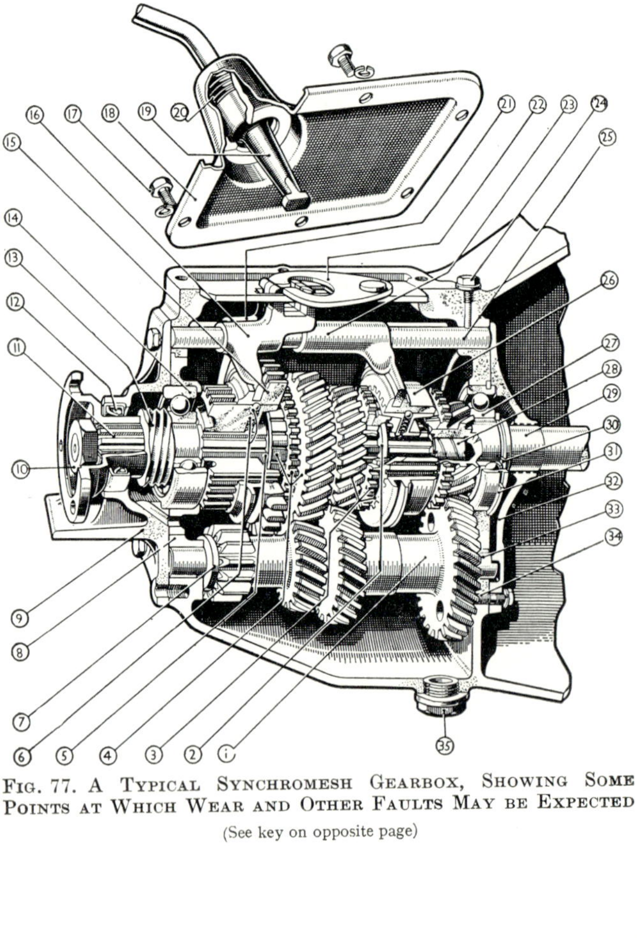

Fig. 77. A Typical Synchromesh Gearbox, Showing Some Points at Which Wear and Other Faults May be Expected

(See key on opposite page)

must be in the same plane ; otherwise a rhythmic variation in speed will take place at the yoke attached to the rear axle driving flange. When correctly assembled the arrows stamped on the splined portion and on the shaft will be in line.

Check the Propeller Shaft. When the universal joints are in good condition, a bent propeller shaft must be suspected. Lack of truth amounting to only three or four thousandths of an inch at the centre of the shaft is sufficient to cause vibration at high speeds. The shaft should be tested for straightness, therefore, by means of vee-blocks and dial gauge.

Possible Rear Axle Troubles. The rear axle is less likely to cause vibration, although serious damage to the driving gears or differential could cause the trouble ; such defects, however, would be almost certainly accompanied by noise that could not be overlooked. A more likely fault is an unbalanced rear wheel or tyre ; this can be checked as indicated in Chapter VIII. Remember, too, that tyre noise can be mistaken for back-axle hum ; when diagnosing rear axle faults, therefore, first eliminate this possibility by driving on an unmade road or a similar surface to ascertain whether the noise disappears under these conditions.

If whine or hum is present in a worn axle, it is seldom

Key to Fig. 77

1. Layshaft gear cluster : gears worn or chipped
2. Front thrust washer worn
3. Third speed wheel teeth worn ; synchronizing cone and coupling damaged ; bush worn
4. Second speed wheel ; faults as with third speed assembly
5. Rear thrust washer worn
6. Synchronizing springs weak ; balls worn
7. Layshaft thrust washer worn
8. Layshaft rear bearing worn
9. Rear cover not tightly bolted up
10. Third motion shaft nut loose
11. Splines on third motion shaft and coupling worn
12. Rear oil seal defective
13. Speedometer driving gear worn or damaged
14. Rear ball bearing worn, balls and races pitted, or tracks poor fit in housing or on shaft
15. First speed wheel teeth chipped
16. First and second speed selector fork damaged
17. Top cover screws loose
18. Top cover distorted ; gasket defective
19. Gear lever bent or fractured
20. Gear lever spring weak or fractured
21. Reverse fork bent or fractured
22. Gear change gate with interlocking arm damaged
23. Third and fourth speed fork bent or fractured
24. Fork rod locking pin loose
25. Forward speeds selector rod bent
26. Third and fourth speed coupling sleeve binding or teeth damaged
27. Front ball bearing worn ; tracks and balls pitted ; races loose in housing or on shaft
28. First motion shaft bush worn
29. First motion shaft worn
30. Locking ring for first motion shaft displaced
31. Spring locking ring for front ball bearing displaced
32. Front cover loose ; oil leakage
33. Front layshaft bearing worn
34. Front cover screws loose
35. Oil drain plug loose ; washer omitted or defective

possible to cure it by re-adjustment of the mesh of the crown wheel and pinion. Normally, the reference to the effect of incorrect mesh of the gears in Chart No. 15 applies only when new gears have been fitted. If the mesh of the gears has been

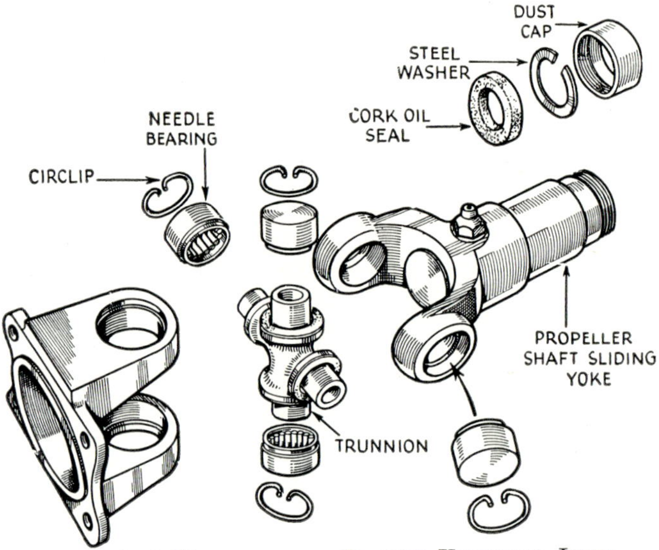

Fig. 78. A Needle-roller Bearing Universal Joint Dismantled

In addition to the bearings and oil seals, the splines inside the sliding yoke should be examined for wear

altered by increased end-float on the driving pinion, however, caused by a faulty pinion bearing or the pinion adjustment becoming distorted for some reason, it may be possible to cure axle hum by restoring the correct mesh of the gears, provided that they have not been allowed to run incorrectly meshed for an appreciable time.

Detecting Backlash. The detection of backlash in the transmission is fairly straightforward when an open propeller shaft is fitted. Jack up one rear wheel, and hold the rear axle driving flange stationary while an assistant rotates the road wheel lightly in each direction. Any excessive backlash

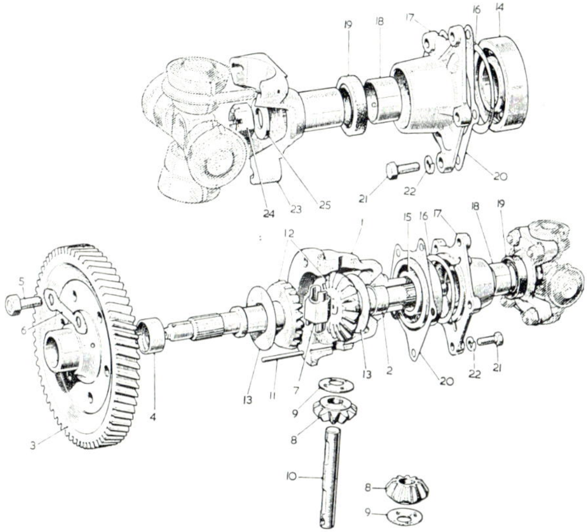

FIG. 79. DIFFERENTIAL AND FINAL-DRIVE GEARS OF B.M.C. 1100 RANGE

1. Differential case
2. Case bush
3. Drive gear
4. Gear bush
5. Gear bolt
6. Lock washer
7. Thrust block
8. Differential pinion
9. Pinion thrust washer
10. Centre pin
11. Pin peg
12. Differential gear
13. Gear thrust washer
14. Drive gear bearing
15. Case bearing
16. Bearing shim
17. End cover
18. Cover bush
19. Oil seal
20. Cover joint
21. End cover screw
22. Washer
23. Driving flange
24. Flange nut
25. Washer

CHART No. 14

PROPELLER SHAFT AND UNIVERSAL JOINT TROUBLES

- **Vibration**
  - Check whether vibration is caused by units other than propeller shaft: *e.g.* engine, clutch, gearbox and rear axle (see text)
    - If localized in propeller shaft
      - Open Propeller Shafts
        - Worn bearings in universal joints.
        - Worn splines on sliding portion.
        - Distorted or torn fabric universal joints.
        - Centering plate in fabric joint incorrectly fitted or omitted.
        - Worn or perished rubber bushes in universal joint.
        - Bent propeller shaft.
        - Incorrect assembly of front yoke on splines (see text).
        - Worn centre steady bearing.
        - Transmission vibration damper discs worn, greasy or binding.
      - Enclosed Shafts and Torque Tubes
        - Worn shaft bearings.
        - Torque tube ball joint worn.
        - Torque tube anchorage incorrectly adjusted.
        - Worn universal joints.
        - Worn splines on shaft.
        - Yokes incorrectly assembled
- **Noise**
  - Usually knocking when load in shaft is reversed
    - Check that knocks or rattles are not due to faulty clutch, gearbox or rear axle. (Charts Nos. 11, 12 and 15)
      - Worn bearings in universal joints.
      - Worn splines.
      - Incorrectly adjusted torque tube anchorage.
- **Oil Leakage**
  - Open Propeller Shafts
    - Overlubrication of splined portion.
    - Defective oil seal in splined portion.
    - Defective oil seals in universal joint bearings.
  - Enclosed Shafts and Torque Tubes
    - External leakage
      - Defective oil seal at lip of ball joint.
      - Defective oil seal in sliding sleeve. or gland requires adjustment.
      - Air vent clogged.
    - Internal leakage into rear axle
      - Defective oil seal between propeller shaft and torque tube.
      - Air vent clogged.
      - Oil return passage to gearbox obstructed.
- **Excessive Backlash**
  - Verify that backlash is confined to propeller shaft (see text)
    - Worn universal joint bearings.
    - Worn splines in sliding joint.
    - Flange bolts on universal joints loose and/or holes worn oval.

CHART No. 15

REAR AXLE FAULTS

- **Noise** — First verify that noise is caused by rear axle (see text)
  - **Whine or Hum**
    - On drive
      - Worn gears.
      - Pinion meshed too deeply with crown wheel (see text).
    - On over-run
      - Worn gears.
      - Pinion not meshed deeply enough with crown wheel (see text).
    - At constant speed
      - Lack of lubricant.
      - Worn gears.
      - Worn races and thrust bearings.
    - When turning corners
      - Worn differential gears.
  - **Knock or Click**
    - Chipped gear teeth.
    - End-float in worm (worm-drive axles).
    - Foreign matter in axle casing.
    - Worn axle shaft splines.
    - Worn wheel hub keyway and key (Chart No. 16).
    - Loose wheel hub (Chart No. 16).
- **Oil Leakage** — Check oil level. If level rises although oil is not added to axle, see "Enclosed Shafts and Torque Tubes," Chart No. 14
  - Check for signs of external leakage
    - Leakage at flanges of drive gear and differential housing
      - Flange bolts loose.
      - Defective gasket.
      - Flange distorted.
    - Leakage into brake drums
      - Defective oil seals in hub or axle.
      - Air vent clogged.
- **Excessive Backlash** — Verify that backlash occurs in rear axle (see text)
  - Worn bevel gear and pinion.
  - Worn worm and wheel (worm-drive axles).
  - Incorrectly meshed gears or worm and wheel.
  - End-float in worm.
  - Worn axle shaft splines, keys or keyways (Chart No. 16).
  - Loose gear cage.
- **Drive not Transmitted** — Verify that fault is in axle (see text)
  - Broken axle shaft.
  - Chipped or stripped gear teeth.
  - Differential parts fractured.
  - Axle shaft splines or key fractured Chart No. 16).

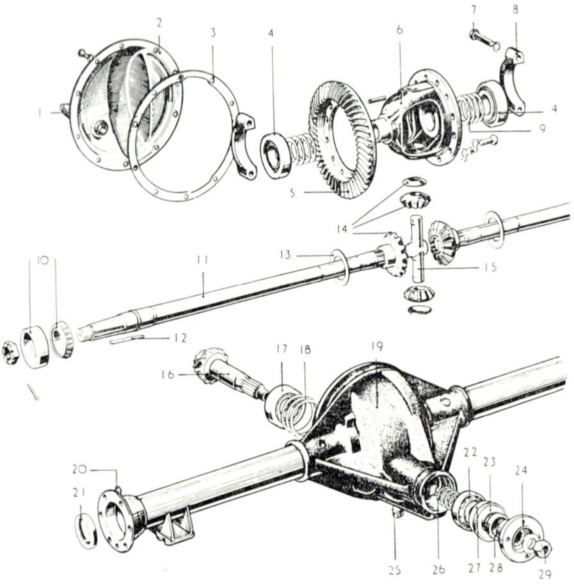

FIG. 80. TYPICAL SALISBURY REAR AXLE, SHOWING SOME POSSIBLE FAULTS

1. Loose drain plug
2. Distorted cover flange
3. Sealing ring leaking
4. Worn differential bearing
5. Drive gear worn or incorrectly adjusted
6. Differential case damaged
7. Differential cap screw loose or broken
8. Bearing cap loose
9. Incorrect number of shims fitted
10. Axle shaft bearing worn
11. Axle shaft worn or broken
12. Key worn or broken
13. Side gear thrust washer worn
14. Differential gears and pinions worn or chipped
15. Differential pinion shaft worn
16. Drive pinion worn or incorrectly adjusted
17. Inner pinion bearing worn
18. Incorrect pinion bearing shims fitted
19. Gear case and axle tube assembly damaged
20. Lubrication nipple choked
21. Shaft oil seal leaking
22. Outer pinion bearing worn
23. Pinion oil seal leaking
24. Driving flange loose on splines
25. Drain plug loose (if fitted)
26. Incorrect pinion bearing shims fitted
27. Pinion oil slinger damaged
28. Dust shield damaged
29. Pinion shaft nut loose

can be attributed to some or all of the hub, axle shaft, and rear axle faults shown in Chart No. 15.

Next, with the rear axle and gearbox driving flanges held stationary, the play in the universal joints and splined joint can be assessed. Finally, by rotating the driving flange of the gearbox with the gears engaged in turn, the wear on the gears, dogs, and the clutch hub splines can be ascertained. It is possible to isolate backlash in the gearbox from play on the clutch only when it is mounted separately from the engine and driven by a cardan shaft, an unusual arrangement nowadays.

CHAPTER VIII

THE WHEELS, HUBS, AND TYRES

THE faults dealt with in Charts Nos. 16 and 17 are bound up to a considerable extent with the steering system troubles covered in the next chapter and, to a lesser extent, with braking system faults. Wheel-wobble and bounce, for instance, will affect the steering; oil-leakage into the brake drums will reduce braking efficiency and worn hub bearings may cause brake judder; tyre condition will affect both braking and steering. Conversely, steering alinement faults and uneven braking will be reflected in rapid and uneven tyre wear.

Testing and Adjusting the Front Hub Bearings. The first test referred to on Chart No. 16 aims at checking the degree of "shake" in the front hub bearings. With the wheel jacked up, grasp the tyre at top and bottom and attempt to rock the wheel in a vertical plane about the hub. Any movement detected will be a combination of hub-bearing wear and wear on the steering swivel pins and bushes. To eliminate the latter, a wooden wedge should be forced between the stub axle fork and the end of the axle beam, or the axle carrier on independent front suspension systems.

If appreciable "shake" can still be felt with the wedge in place, the bearings may be excessively worn: on the other hand, when taper-roller races are fitted, a definite end-float of between 0·005 in. and 0·020 in. may be specified by the manufacturer. Adjustable ball bearings of the cup-and-cone type, however, should usually be adjusted so that the bearings are just free to rotate without binding, but have no end-float. The safest plan, therefore, is to ascertain the manufacturer's recommendation whenever possible.

Adjustment of the bearings is carried out by removing the wheel, taking off the hub cap (if any) and extracting the split-pin which locks the hub nut. This nut should then be fully tightened and then slackened back by one castellation to allow the bearing to rotate freely, but without "shake." When a definite end-float is specified this should be measured with the aid of a dial gauge, fitted to a clamp or bracket

CHART No. 16

WHEEL, HUB AND BEARING TROUBLES

Wheel Wobble — (Trouble confined to individual wheels, as distinct from wobbling of front wheels in unison, as described under steering faults in Chart No. 18) — Test for wear of hub bearings as described in text
- Wheel can be rocked
  - Hub bearings worn.
  - Hub bearings incorrectly adjusted
  - Hub nut loose.
- No bearing wear evident — Test wheel and tyre for truth (see text) — If run-out is excessive
  - Wheel buckled.
  - Tyre incorrectly fitted.
  - Tyre casing distorted.
  - Axle driving shaft bent (on driving wheel).
  - Hub flange distorted.

Wheel-Bounce — Test for eccentricity of wheel and tyre (see text)
- Eccentricity evident
  - Tyre incorrectly fitted.
  - Tyre bulged.
  - Wheel damaged or spokes incorrectly tensioned on wire wheel.
- No eccentricity
  - Tyre pressure incorrect.
  - Wheel and tyre out of balance (see text).
  - Unbalanced brake drum.
  - Suspension defects (see Chart No. 23).

Wheel Loose on Hub — Check tightness of wheel nuts
- If tight
  - Nuts bottoming on thread of stud or end of stud bottoming in acorn-type nut.
- If nuts become loose during running
  - Grit between mating surfaces of nut and wheel.
  - Grit between wheel and hub.
  - Grit on cones of Rudge-Whitworth (knock-off) wheel.
  - Nuts changed over to wrong side of car when left-hand and right-hand threads are used to give self-tightening effect.

Hub Loose on Driving Shaft — Check that backlash is between hub and shaft, not elsewhere in transmission (see text) — If diagnosis is confirmed
- Loose hub nut.
- Worn key and keyways in hub and on shaft.
- Worn splines in hub and on shaft.

Oil or Grease Leakage into Brake Drums — Trouble may be due to over-lubrication or use of too thin a grade of oil or grease — If correctly lubricated
- Defective oil seals in hubs.
- Defective oil seal on axle shaft (driven hubs).
- Oil level in driving axle casing too high.
- Oil vent in driving axle casing clogged.
- Fault may be due not to hubs but to over-lubrication of brake shoe pivots or operating spindle.

attached to the brake drum or a wheel stud, so that its plunger rests on the end of the stub axle. Pushing and pulling on the brake drum will then indicate the amount of end-float on the

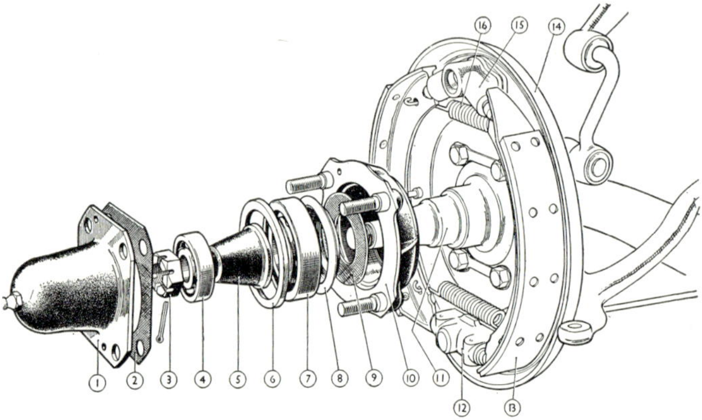

Fig. 81. A Front Hub Dismantled, To Show the Component Parts and the Location of Possible Faults

1. Hub cover damaged or flanges distorted
2. Joint washer defective
3. Bearing adjusting nut incorrectly adjusted ; cotter pin omitted
4. Outer ball bearing worn ; balls and tracks pitted ; races poor fit in housing or on shaft
5. Distance piece
6. Incorrect thickness of packing washer
7. Inner bearing ; defects similar to outer bearing (see 4)
8. Incorrect thickness of distance ring
9. Defective oil retainer
10. Hub flange distorted ; studs loose or stretched
11. Stub axle bent ; bearing seatings worn
12. Brake expander : cones, rollers and tappets defective
13. Brake shoe ; friction lining worn or glazed ; rivets loose ; shoe distorted or fractured
14. Brake backplate bent or loose on axle
15. Brake adjuster : incorrect adjustment ; seized adjuster spindle
16. Weak or broken brake shoe pull-off springs

bearing. Needless to say, the brake drum securing screws must be fully tightened and it should be verified that the brakes are not binding before making the test.

When non-adjustable bearings are fitted, renewal is, of course, the only cure for excessive wear. This applies also to the rear hub bearings, which are seldom adjustable.

Testing a Wheel for Trouble. When the condition of the hub bearings has been checked, the wheel should be tested to determine whether it is buckled or eccentric. These faults are common on older cars fitted with wire-spoked wheels, and are caused by kerb collisions and stretched, or broken spokes. Being more robust, the modern all-steel wheel is less likely to be distorted, but this fault cannot be altogether ruled out. A more usual trouble, which will have just the same adverse effect on steering and tyre wear, is an incorrectly fitted or distorted tyre.

To check the wheel and tyre is a simple matter. With the wheel jacked up, rest a piece of chalk on a firm support at hub-height so that it just touches the outside of the wheel rim. When the wheel is rotated the chalk will mark the points of maximum run-out. Next, make a test with the chalk on the inside flange of the rim : this will show up any eccentricity.

Having checked the wheel, repeat the test with the chalk against the tyre wall at the base of the tread to reveal incorrect fitting of the tyre or a bulged or distorted cover. Finally, test with the chalk on the centre line of the surface of the tread to show up any eccentricity caused by the tyre being incorrectly fitted or due to uneven tread wear.

Correcting any Faults Revealed. Truing a buckled or eccentric wheel is a job for an experienced wheel-builder. A wire-spoked wheel is brought into correct alinement by tightening the appropriate spokes and loosening others. Although this sounds simple in theory, and can, in fact, be carried out without difficulty by an expert, the novice who attempts the operation is likely to make matters worse rather than better. A pressed-steel wheel can be straightened only in a press. If the run-out or eccentricity on this type of wheel does not exceed $\frac{3}{16}$ in., however, truing is not necessary.

When a tyre is found to be incorrectly fitted it is, of course a simple matter to remove and refit it correctly. Most tyres have a line moulded on the side-wall close to the bead. This line should be parallel to the rim when the tyre is correctly fitted. If run-out or eccentricity is caused by tread or casing defects, however, the only cure is to fit a new tyre. If a replacement is not available, the tyre should be fitted to a rear wheel ; the effect of the faults will be less noticeable than on the front wheels, where run-out may cause wheel-wobble, and eccentricity may set up wheel " tramp."

Other Causes of Wheel-Wobble. If a wheel and tyre appear to be in good condition, and wobble is experienced, the wheel should be tested on another hub. If the trouble now disappears, the original hub flange should be tested in a manner similar to that adopted when checking the wheel. A buckled hub flange is less common than a buckled wheel, but can be caused by accidental damage. In the rear wheels, wobble can sometimes be traced to a bent axle driving shaft

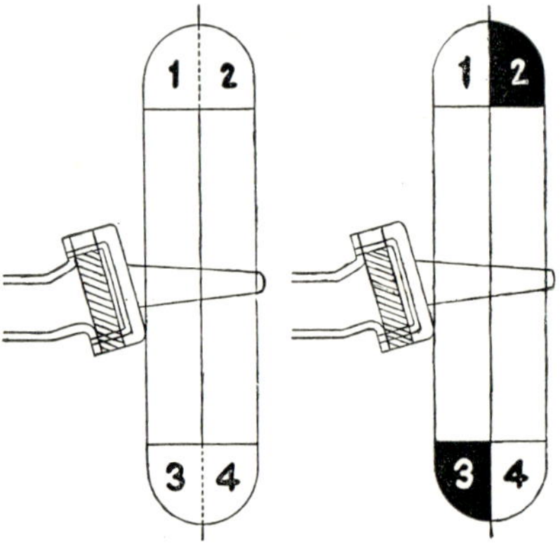

FIG. 82. THE WHEEL ON THE LEFT IS BOTH STATICALLY AND DYNAMICALLY BALANCED. THAT ON THE RIGHT IS STATICALLY BALANCED BUT DYNAMICALLY UNBALANCED, SINCE THE HEAVY MASSES, 2 AND 3, ARE ON OPPOSITE EDGES OF THE TYRE

when the shaft is rigidly attached to the hub or is integral with it, as on semi-floating and three-quarter-floating axles. The only cure is to remove the hub and shaft and either to fit a new shaft, or to straighten the shaft in a press if it is only slightly bent, free from cracks, and otherwise undamaged.

Checking Wheel Balance. Unbalanced front wheels and tyres are a common cause of steering troubles such as wheel-wobble and tramp, while unbalanced rear wheels will cause vibration which may be attributed in error to transmission faults.

On older models an appreciable degree of unbalance may sometimes exist without giving rise to noticeable symptoms but most modern cars are very sensitive to wheel balance and even the amount of unbalance which may be present in a new tyre may be sufficient to cause noticeable vibration, usually in the 60–70 m.p.h. range. The wheels should therefore be checked for balance at 6,000-mile intervals and also whenever the tyres are removed and refitted. With modern equipment it is possible to check each wheel without removing it from the car. This is obviously the most satisfactory method, since any unbalance in the brake drum and disc or hub assembly will be corrected at the same time.

Lack of balance can generally be traced to the presence of patches or gaiters in the tyre, uneven tread wear and, of course, to the weight of the tyre valve itself. Some degree of unbalance is usual, even on new covers. Often a spot on the side of the tyre marks the lightest point, so that the tyre can be fitted with the spot adjacent to the tyre valve to equalize the balance as far as possible.

What is Dynamic Balance? The most satisfactory method of balancing wheels and tyres is to use a properly designed balancing machine, preferably of the type which enables both static and dynamic balance to be checked. The latter term refers to the balance of the tyre when rotating at speed, a factor which cannot be checked without the correct equipment. A wheel is dynamically unbalanced when the heavier points are towards the inside or outside of the rim or tyre tread, instead of on the centre line of the wheel and tyre.

Consider what happens in a front wheel : when the wheel is rotating it will attempt to turn the stub axle until a line drawn through the heavier points is at right angles to the axis of rotation of the wheel. When a heavier point is on the inside of the tyre it will swing the wheel towards the right when the unbalanced portion is in front of the axle, and will exert a turning force to the left when it moves to behind the axle as the tyre rotates. Fig. 82 helps to make this clear. Consequently, as the wheel rotates it tends to swing to the right and left alternately, and unless this tendency is resisted by the steering gear, wheel-wobble or "shimmy" is set up. Since the method of balancing a wheel on a dynamic balancing machine depends on the design of the machine, the manufacturer's advice must be followed.

How to Check Static Balance. By static balance is meant the tendency of an unbalanced wheel always to come to a standstill with the heavy portion at the lowest point, if the wheel is jacked up clear of the ground and is fitted to a freely rotating hub. The centrifugal force exerted by an unbalanced mass when the wheel is rotating at speed causes the wheel to " patter " or " tramp."

It is a simple matter to check a wheel for static balance by fitting it to a front hub, first making sure that the hub bearings are in good condition and that the hub rotates freely. For

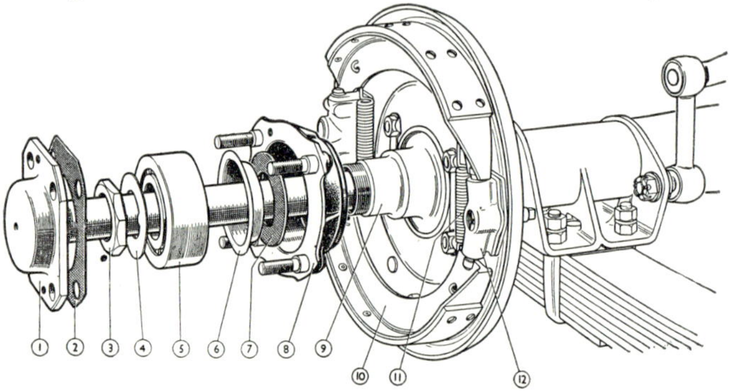

FIG. 83. TYPICAL REAR HUB DISMANTLED. BELOW ARE SOME OF THE FAULTS TO LOOK FOR WHEN INSPECTING THE HUB

1. Axle shaft flange distorted
2. Defective joint washer
3. Hub lock nut loose
4. Locking washer not secured
5. Double ball race : balls and tracks worn and pitted ; inner and outer races poor fit on shaft and in housing
6. Distance ring
7. Defective oil retainer
8. Hub flange distorted. Studs bent or stretched
9. Axle tube damaged
10. Brake shoe
11. Brake pull-off spring
12. Brake expander

accurate results it is best to eliminate all friction : dismantle the hub, remove the oil seal, wash the grease from the bearings with petrol, and reassemble them with a light oil as a lubricant, being careful not to over-tighten the hub retaining nut.

Rotate the wheel and allow it to come to rest. Mark the lowest point with chalk, and repeat the test several times. If the chalk mark is not at the lowest point of the tyre on each occasion, either the wheel and tyre are in perfect balance or there is friction in the hub. When the chalk mark indicates the heaviest point of the wheel and tyre, a weight must be attached to the rim at a point opposite to the chalk mark, altering the amount and position of the weight until satisfactory balance is obtained.

On some sports cars lead balancing washers are fitted to spindles beneath metal caps at three points on the rim of the wheel. When pressed-steel wheels are fitted, clip-on balancing weights can sometimes be obtained from the car manufacturer. If these are not available it will be necessary to drill holes in the rim at three positions, 120 deg. apart, so that balancing washers may be attached by bolts, which must, of course, be fitted in such a manner that there is no possibility of damaging the tyre beads or the inner tube. The best position is usually in the well of the rim. Coach bolts should be used, their domed heads being covered by the rim tape. On wire-spoked wheels it is a simple matter to wrap lead wire around appropriate spokes, securing it in place by a wrapping of insulating tape.

Examining Hubs and Bearings for Faults. When the various tests given in Chart No. 16 indicate hub and bearing faults the hubs must be removed in order to inspect the parts. It may be possible to tap a front hub off the stub axle, but in many cases a hub extractor must be used. The car manufacturer can usually supply a special tool when one is required.

Similarly, the removal of the rear hubs may call for the use of an extractor. When the hubs are integral with the axle shafts, however, it is sometimes necessary to remove the inspection cover from the rear axle in order that circlip retainers on the inner ends of the shafts may be removed.

When the hubs have been removed, check the bearings for pitting, cracking, wear of the races, and chipped or broken balls. The fit of the races in the axles and in their housings must also be checked, and the races and housings examined for scoring or " fretting." The rule for bearing fits is that the race which fits on or in a rotating member must be an interference fit, while that in the stationary member should be a light push fit or " creep " fit, to allow it to align itself with the rotating race.

The oil seals should be examined and renewed if necessary. They may take the form of felt rings; or leather cup washer seals, the lips of which are held lightly in contact with the axle by coil springs, may be used. On rear axles a cork gland may be cemented into the axle casing; check this for wear or breaking-up.

The hub flanges should be tested for distortion as already described, and the wheel studs carefully examined for signs of stretching, bending, or looseness in the hub. On the rear hubs check the condition of the splines, or key and keyway

CHART No. 17

CAUSES OF TYRE TROUBLES

Trouble		Symptom	Cause
Excessive or Uneven Tread Wear	Appearance of tread will give good indication of cause of wear	Wear on both sides of tread; centre less worn	Low tyre pressure—check pressure weekly (see "Loss of Pressure" below). Tyre overloaded.
		Wear on one side of tread	Incorrect camber angle (see Chapter 9). Driving on steeply cambered roads.
		Wear on centre of tread	Tyre pressure too high.
		"Heel-and-toe" wear on blocks in tread pattern	Heavy braking: generally confined to front tyres as acceleration stresses equalize wear on rear tyres.
		Tread has "filed" appearance, with lips of rubber at edges of pattern	Incorrect alinement of wheels: usually confined to front wheels (see Chapter 9).
		Tread is worn unevenly, in ridges and depressions	Buckled wheel. Wheel and tyre out of balance. Eccentric brake drums or grabbing brakes
		Even, but rapid wear	Fast driving. Rough road surfaces.
Loss of Pressure		Loss of air does not exceed 1–5 lb. per week	Normal diffusion of oxygen from tube.
		Loss exceeds 1–5 lb. per week	Tubeless tyre incorrectly fitted. Tube punctured. Valve faulty. Porous tube—or tubeless cover. Leaking patch on tube—or cover.

Cover Fractures and Bursts

- Pressure too low
  - Under-inflation causes excessive flexing of tyre walls, causing fracture of cords in cover
  - Overheating of cover due to excessive flexing of plies.
  - Kerb collision causes fracture of plies in cover.
- Pressure too high
  - Excessive pressure may cause blow-out at weak spot in cover caused by previous damage or faulty repair.
- Pressure correct
  - Kerb collisions.
  - Tread or side wall cut by sharp object.
  - Neglected cuts in tread allow water and grit to enter and rot fabric plies.

Faults on Repaired Covers

(When major or minor damage described above has been repaired by cutting out damaged tread and plies of cover, and vulcanizing patch and rubber filling in place)

- Rubber filling plug becomes detached — Test rubber of plug by pressing blunt probe into surface
  - If mark remains, rubber is not properly cured
    - Insufficient curing time.
    - Curing temperature too low.
  - Surface of rubber springs back into place
    - Bad preparation of surfaces before vulcanizing.
    - Insufficient vulcanizing pressure.
    - Angle of bevel on plug incorrect.
    - Incomplete removal of original damage.
- Cover bursts at site of repair — Assuming that plug defects above do not exist, burst is due to weak reinforcing patch:
  - Insufficient number of plies in reinforcing patch.
  - Incomplete removal of damage.
- Rapid wear of replaced section of tread
  - Tread compound insufficiently cured.
  - Poor quality tread compound.
  - Tread pattern not properly reproduced to match remainder of tread.

by which the hub is attached to the axle shaft. The slightest slackness will lead to rapid wear, owing to the impact loads on the surfaces during acceleration and over-running.

On centre-lock or " knock-off " hubs, the splines on which the wheel slides must also be checked. It is a mistake, however, to assume that these splines transmit the drive to the wheel. This is, or should be, performed by the coned surfaces of the hub and wheel, which should therefore be free from corrosion or " fretting." If the wheel has been run with the retaining nut loose, not only will the cones be fretted but the splines will also be worn.

Diagnosing Tyre Troubles. Little need be said in amplification of Chart No. 17, which covers the causes of any tyre troubles likely to be experienced. It will be seen that the most usual cause of rapid tyre wear and damaged tyres is incorrect pressure. It is not always realized that a loss of pressure of from one to five lb per sq. in. per week is quite a normal occurrence when a normal rubber inner tube is fitted. It is due to a process termed "diffusion," in which the rubber of the tube absorbs oxygen from the compressed air in the tyre ; as the rubber now contains excess oxygen a corresponding amount is given off from the outer surface of the tube. There is thus a slight loss of pressure caused by the gradual decrease in the oxygen content of the air in the tube. Synthetic-rubber tubes and tubeless tyres do not—or should not—suffer from this trouble.

Another faulty diagnosis concerns the cause of sudden deflation of a tyre. This may, of course, be caused by a major defect, such as a burst or " blow out " in the cover, or by the entry of a large, sharp object which causes extensive damage to the cover and tube. Sudden deflation should not, however, be attributed to a " puncture " in the normal sense of the word. When a nail or other sharp object passes through the tyre and pierces the tube, little or no loss of air occurs at first, since the pressure of air in the tube seals the lips of the rubber around the object which penetrates the tube.

As the tyre flexes during normal driving, however, the relative movement between the lips of the puncture and the object which caused it results in the edges of the puncture becoming frayed and the hole in the tube becoming enlarged. A slight loss of pressure then occurs, but the tyre may run for days, or even weeks, with no other sign that anything is amiss, particularly a tubeless tyre. The regular loss of pressure is a danger signal which should be recognized, however, when the tyre pressures are tested at regular intervals. Provided that the pressure loss cannot be traced to a faulty tyre valve, a loss of pressure exceeding five pounds per week indicates that a "puncture" is likely to occur in the near future.

Mention of the tyre valve calls for a reminder that the spring-loaded air seal in the valve is provided only as a convenient means of testing the pressure and inflating or deflating the tube. It is not intended to function as a permanent air seal. Moisture, grit or ice can quickly cause the valve to leak. The screwed valve cap is therefore provided by the manufacturer as a positive seal for normal running. Make sure that the caps are in place, screwed tightly home, and that the rubber sealing washer in each cap is in good condition.

CHAPTER IX

STEERING GEAR AND FRONT SUSPENSION DEFECTS

SINCE the first symptoms of steering gear and front suspension faults usually manifest themselves in the form of wheel-wobble, heavy steering, wandering, or a tendency to pull to one side, Chart No. 18 is devoted to analysing these troubles. It will be seen that although wear in the steering gearbox or steering connexions is usually blamed for these troubles, the cause can, and often does, lie in a number of other components.

It should also be remembered that some steering defects, such as oversteer and understeer, can be attributed to incorrect tyre pressures. On front-wheel drive or rear-engined, rear-wheel drive models, for example, the recommended pressures for the front and rear tyres may differ by as much as 10 lb per sq in., owing to the heavy load carried by the driving wheels when the engine and transmission unit is mounted above them, and if this pressure differential is not maintained the car will not steer and handle properly.

Generally speaking, a car which has a tendency to understeer—that is, to run wide on corners, calling for considerable effort at the steering wheel to maintain the correct course—can be cured either by raising the pressures in the front tyres or lowering those at the rear. Conversely, oversteering—the tendency of the rear of the car to swing out on corners so that it is necessary to "unwind" the steering wheel part-way through the turn—can be cured or reduced by raising the rear tyre pressures or lowering the pressures at the front.

It must be remembered, however, that both these faults may be to some extent inherent in the design of the car. Many front-wheel-drive models tend to understeer when accelerating around the corner and to oversteer if the throttle is suddenly closed while cornering. Rear-engined models, owing to the preponderence of weight at the rear, tend to be unstable when cornering fast, when the tail tends to be swung outwards and there may be a tendency to sudden breakaway which may be difficult to control. Clever design can often minimize either of the faults just described but the characteristics of the particular car must be borne in mind when attempting to diagnose complaints of poor steering.

Checking the Steering Mechanism for Wear. It is a simple matter to check the various parts of the steering mechanism for wear and to isolate the trouble in one or more components. Jack-up the front axle, and begin by grasping a front wheel at the top and bottom of the tyre and attempting to rock it vertically about the stub axle. Any appreciable movement will indicate wear of the swivel pin and bushes. Slackness of the wheel bearings will also contribute to the total movement, however, so that these should be checked and adjusted as described in Chapter VIII.

Next attempt to lift the wheel vertically, thus revealing any wear on the thrust washer or bearing which normally carries the weight of the front end of the car. Vertical movement caused by wear of the thrust bearing is less serious than swivel pin wear, but should be put right while the swivel pins and bushes are being renewed.

Attention can now be turned to the steering linkage and steering gearbox. While the end of the drop arm is held stationary by an assistant, rock each wheel about its swivel pin while watching the ball joints in the track rod and drag link for slackness. While the drop arm is prevented from moving, determine the amount of backlash in the steering gearbox by rotating the steering wheel lightly in each direction. If the free movement exceeds about two inches on the rim of the wheel the steering gearbox will need attention. First make sure however, that the drop arm is not loose on the splines attaching it to the steering gearbox shaft, or that the cotter-bolt is not slack or worn when this method of attachment is used. Another likely fault to look for is looseness of the bolts attaching the steering gearbox to the chassis, or a loose steering column bracket.

On the majority of modern cars self-adjusting joints are fitted to the steering gear and front suspension swivels. Rubber, nylon or polytetrafluoroethylene (p.t.f.e.) bushes, seatings and thrust washers render regular lubrication unnecessary. When checking for stiffness or squeaks in the steering or for the cause of undue slackness, however, possible lack of lubrication should not be overlooked. Sometimes blanking plugs are fitted to the joints and these must be

CHART No. 18

DIAGNOSING STEERING FAULTS

The faults listed below are those which are revealed by steering troubles during normal driving. To diagnose faults in specific components, refer to Charts Nos. 16, 17, 18, 19, 20, 21 and 22.

EXCESSIVE BACKLASH AT STEERING WHEEL: STEERING INDEFINITE
→ Most probable cause is wear of steering ball joints and steering gearbox. Check as in text and see Chart No. 19
→ Test swivel pins and bushes for wear (see text). Check wheel bearings for slackness (see Chapter VIII)
→ If no excessive slackness present, check for:
- Steering gearbox mechanism worn or incorrectly adjusted (Chart No. 19).
- Steering connexions worn or too loosely adjusted (Chart No. 19).
- Steering gearbox loose in chassis.
- Steering drop arm loose on splines or key.

CAR WANDERS
→ Check for excessive backlash in steering and wear on swivel pins, bushes and wheel bearings—(see above
→ If no excessive backlash or wear is present:
- Low or even tyre pressures.
- Rear tyre pressures lower than front.
- Car overloaded at rear.
- Binding at some point in steering linkage due to over-adjustment or lack of lubrication.
- Insufficient castor angle or reverse castor (Chart No. 19).
- Front or rear axle displaced on spring or securing bolts loose.

WHEEL WOBBLE OR "SHIMMY"
→ At high speed ("shimmy")
→ Check wheel balance (Chapter VIII)
- Wheels dynamically unbalanced:
  - Tyres incorrectly fitted (Chart No. 17).
  - Uneven wear of tyre (Chart No. 17).
  - Incorrect balance weights fitted.
- Wheels correctly balanced:
  - Wheel rims buckled.
  - Worn steering connexions.
  - Incorrect toe-in of front wheels (Chart No. 19).
  - Incorrect castor action (Chart No. 19).
  - Incorrect tyre pressures.

- At low speed (wobble)
  - As for high speed "shimmy," but check following points particularly:
    - Tyre pressures incorrect.
    - Castor angle insufficient—probable cause is flattened front springs.
    - Excessive backlash in steering system (see above).

Car Pulls to one Side

- Under normal driving conditions
  - Uneven tyre pressures.
  - Incorrect toe-in of front wheels (Chart No. 19).
  - Castor or camber angles differ on front wheels (Chart No. 19).
  - Front or rear axle slightly displaced on springs.
- When braking
  - Mainly braking faults (see Charts Nos. 21 and 22) but may be aggravated by faults under "Normal driving conditions," above
    - Brakes unevenly adjusted (Chart No. 21).
    - Grease on brake linings (Chart No. 16).
    - One front brake binding (Chart No. 21).

Heavy Steering

- Test with front axle jacked up
  - Steering becomes normal
    - Front tyre pressures too low.
    - Insufficient lubrication of swivel pins and bushes—particularly if worn.
    - Excessive castor angle (Chart No. 19).
    - Chassis alinement incorrect.
  - Steering remains heavy
    - Lack of lubrication of steering gearbox and connexions.
    - Steering gearbox and/or connexions too tightly adjusted.
    - Bent or twisted steering arm—may also foul chassis or axle.
    - Steering column distorted by misalinement of steering gearbox and mountings.

removed at specified intervals to allow lubrication nipples to be substituted and oil or grease to be injected. When no such provision is made, a stiff or squeaky nylon joint can usually be cured by injecting oil.

When checking the steering, particular attention should be paid to the rubber sealing gaiters on the joints and at the ends of a rack-and-pinion gear. A damaged gaiter should be renewed without delay but as there is also the risk that the joint itself may have been damaged by the entry of water or grit, it is a wise precaution to renew this also.

When adjustable joints are fitted, the end plug should be screwed in as far as possible, and then unscrewed just far enough to allow the split pit to be inserted. This should eliminate play without causing the ball to bind. It is as well to test the joint with the wheels turned to full lock in each direction, as the ball may be worn slightly oval owing to the majority of steering movements taking place over only a small arc. If the ball is worn, it may bind after adjustment when the wheels are turned towards full lock. The only cure is to renew the ball, and preferably the cups at the same time.

A dangerous fault with some types of joint is that when the ball becomes worn it may be free to drop out of its socket. As this can cause a disastrous accident, renewal of the ball and cups is a wise policy.

Most of the remaining points in Chart No. 18 are self-explanatory, or are dealt with in Chapters VIII, X, XI or XII. Attention can be turned, therefore, to Charts Nos. 19 and 20, in which the faults diagnosed by the previous chart are dealt with in more detail.

The Importance of Correct Steering Geometry. If a preliminary check-over of the steering mechanism does not reveal any serious faults, or if, after correction of faults, steering is still unsatisfactory, the trouble probably lies in incorrect steering geometry. Without going into the theory of steering geometry in any detail, it can be said that four factors influence the controllability of the car: the toe-in of the front wheels; the castor angle; the combined camber angle and king-pin inclination; and the Ackerman angle. These can be considered briefly in that sequence, which represents their relative order of importance.

Checking the Toe-in of the Front Wheels. Owing to the

fact that the front wheels lean outwards at the top, because of the camber angle imparted to them (as described later) the wheels tend to turn outwards or roll apart when the car is in motion. This is resisted by the track rod, the object of which

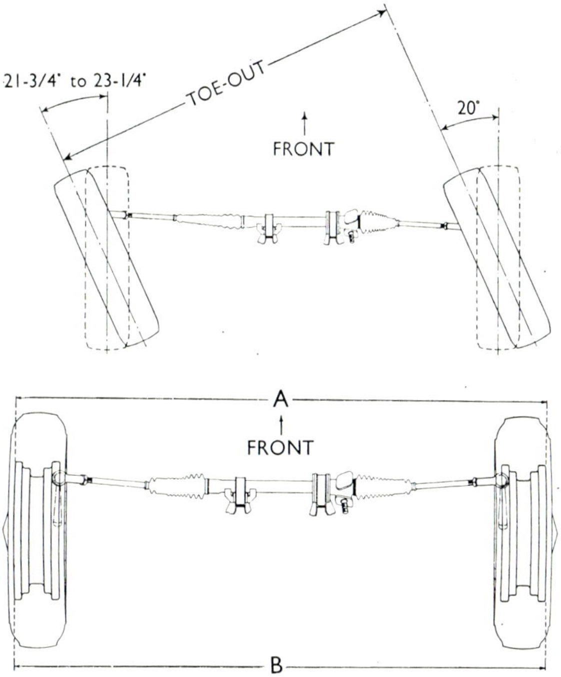

FIG. 84. FRONT-WHEEL ALIGNMENT

Above, the front wheels should normally toe-out on turns. *Below*, if the measurement A is less than B, the wheels toe-in. If it is greater, they toe-out. In both cases the measurement is taken with the wheels straight ahead

is to keep them parallel. As the track rod is slightly elastic, however, it is usual to set the wheels so that they toe-in to a small extent when stationary. The slight compression or extension of the track rod, depending on whether it is fitted behind or in front of the axle, will result in the wheels assuming a parallel position to one another when in motion.

CHART No. 19

CAUSES OF STEERING GEAR FAULTS

- **Incorrect Steering Geometry**
  Good steering will not be obtained unless the essential angles are correct (see text)
  - **Incorrect Toe-in**
    - Incorrectly adjusted track rod.
    - Bent track rod.
    - Bent steering arm on stub axle.
    - Buckled wheel gives incorrect measurement when checking toe-in—adopt method recommended in text.
  - **Incorrect Castor Angle**
    - Front springs flattened.
    - Wedge omitted.
    - Wear of swivel pins and bushes.
    - Incorrect adjustment of upper wishbone bearing on I.F.S. system
    - Incorrect adjustment of radius arms or torque reaction rods.
  - **Incorrect Camber Angle and Swivel Pin Inclination**
    - Worn swivel pins and bushes.
    - Bent stub axle.
    - Bent axle beam or axle carrier.
    - Incorrect adjustment of upper wishbone bearings on I.F.S. systems.
  - **Incorrect Ackerman Angle**
    - Steering arms on stub axle bent.
    - Chassis distorted.
    - Axles displaced on springs.

- **Cam-and-Follower Gears** (Bishop and Ross)
  - Wear of conical roller and tapered groove in cam — Adjust by removing shims from between side cover plate and box or by means of thrust setscrew on high-efficiency gear.
  - Wear on steering column bearing — Replace if felt bush or plain ball bearing. Adjust if adjustable bearing is fitted.
- **Cam-and-Follower Gear** (Marles Weller)
  - Wear between hemispheres in follower and cam track — Replace hemispheres.
- **Hour-Glass Cam-and-Roller Gears** (Marles and Gemmer)
  - End-play on cam — Adjust by removing shims from between bottom cover and box or by threaded sleeve surrounding steering column at top of box.

Backlash in Steering Gearbox The only faults quoted are those which can be cured by external adjustments, which are sufficient for normal service. Other faults entail dismantling the gearbox (see text)	Screw-and Nut Gears (Burman Douglas)	Wear between nut and thread on base of column	No adjustment—renew parts.
		End-float in column	Adjust bearing at top of column—may be necessary to lift steering wheel to reveal nuts.
		End-float in rocker shaft	Adjust setscrew in side cover. Adjustment not fitted on all models.
	Worm-and-Wheel Gears (various makers and designs)	Wear between worm and sector or worm-wheel	Method of adjustment varies: eccentric boss carrying sector; adjustable side plate; provision of complete worm-wheel which can be rotated half-revolution to bring fresh section into use.
		End-play on worm	Adjustment varies: shims beneath bottom cover; adjustable screwed end plug; milled adjusting sleeve at top of box.
		End-float in column	Method of adjustment varies: shims or adjustable ball race.
	Rack-and-Pinion Gears (Various makers and designs)	Wear between teeth of rack and pinion	Rack is usually kept in contact with pinion by spring-loaded plunger.
		Wear on ball joints coupling track rods to rack	Usually spring-loaded; may be adjustable end-plug in rack.
Backlash in Steering Connexions	Self-Adjusting Types	Wear of cups and ball	Some joints can be dismantled and parts renewed, but most modern types must be replaced as complete unit.
	Rubber-Bushed Types	Worn or perished bushes	Bushes can be renewed on some types; otherwise renew complete joint.
	Adjustable Types	Excessive backlash	Adjust screwed end plug. If ball is worn oval, renew ball and cups.

It is essential that the correct alinement should be ascertained for the particular car. The toe-in may vary between $\frac{1}{16}$ in. and $\frac{1}{4}$ in.; some manufacturers specify that the wheels must be exactly parallel; others—a small minority—that they should toe-out slightly. The correct alinement depends on the design of the steering linkage and on the steering angles. Even a slight error is sufficient to cause the car to wander and at the same time to set up extremely rapid tyre wear.

The pointers of the alinement gauge are first adjusted to touch the wheel rims at hub height behind the front axle, the points of contact being marked by chalk spots. The gauge is then removed without disturbing the setting of the pointers, the car is pushed forward until the chalk marks are at hub height in front of the axle, and the gauge is placed in position with one pointer touching the rim at the chalk spot. The distance of the other pointer from the rim is equivalent to the toe-in of the wheels. By using this method and marking the rims with chalk the errors in measurement caused by slightly buckled rims are avoided.

Measuring Castor Angle. The castor angle of the front wheels represents the amount by which the swivel pin is inclined backwards. As a result a line projected through the swivel pin would meet the road ahead of the point of contact of the wheel. The wheel thus, in effect, trails behind its true pivot point, just as the castors on furniture trail behind their pivots and so possess a self-centering action. On a car the castor angle exerts a considerable influence on the stability of the steering, wheel wobble developing if the angle is insufficient, and the steering becoming uncontrollable if a reverse, or negative castor angle exists. The latter state of affairs can exist if wedges, fitted between the axle and the spring to provide adequate castor angle when the springs have flattened, are subsequently refitted the wrong way round, with the thick end towards the front of the car. Wedges are at best a makeshift; the springs should either be reconditioned and re-set to the correct camber, or new springs should be fitted.

On independent front suspension systems castor angle can often be adjusted by means of threaded bushes. If the suspension has been dismantled, it is essential that on reassembly the castor angle should be correctly set. This also applies when the axle is positioned by a radius rod or a torque-resisting rod, adjustment to the length of the rod or the fitting of shims at its anchorage points being employed to vary the castor angle.

Although it is possible to check the castor angle roughly by laying a straight-edge across the axle pad and measuring the height of each end from the ground, and converting the difference in height into degrees by calculation, the only really

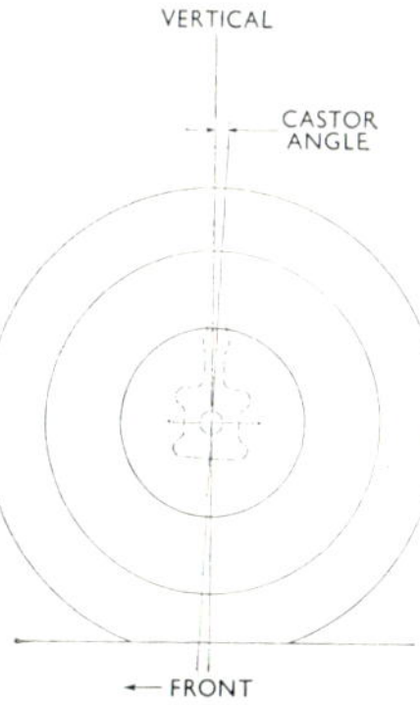

Fig. 85. Castor Angle

accurate method is to use one of the many types of wheel alinement testers available in garages and service stations. In many such gauges the actual castor angle is not measured. Instead the camber angle is measured with the wheel turned through a twenty-degree angle in each direction, and the castor angle, which is calculated from the two camber angle readings, is read off on the scale.

Measuring Camber Angle and Swivel Pin Inclination. These two angles are illustrated in Fig. 86. The camber angle has the effect of reducing steering effort and also ensures that the greater part of the load on the wheel is carried by the inner hub bearing. Swivel pin inclination provides a close approach to centre-point steering, and also helps to keep the stub axles pointing outwards at right angles to the car, thus assisting the castor angle in returning the wheels to the straight-ahead position. If the outward inclination of the wheel is measured

CHART No. 20

POWER STEERING SYSTEM FAULTS

Th faults listed in this chart are confined to the power-assistance system. See also general steering system faults covered in Chart No. 19.

- **LOSS OF POWER ASSISTANCE**
  - Check oil level
    - Oil level correct—Check hoses and connexions
      - If sound, carry out tests under "Check pump drive."
    - Oil level low (may also cause noisy operation)
      - Check system for leaks.
      - Top-up reservoir.
      - Bleed air from system as necessary.
  - Check pump drive
    - Drive faulty
      - Belt broken or slack.
      - Pulley damaged or loose on shaft.
    - Drive correct
      - Carry out tests under "Check pump relief and flow-control valves."
  - Check pump relief and flow-control valve
    - Valve sticking
      - Clean, remove burrs if necessary or renew.
    - Valve in good condition—Check pump for wear
      - Pump sound—Check power cylinder.
      - Pump worn—Recondition or renew.
  - Check power cylinder
    - Power cylinder in good condition
      - Earlier stages of diagnosis faulty.
    - Power cylinder worn or damaged
      - Recondition or renew.

as shown in the drawing, the apparent angle will naturally be affected by both the swivel pin inclination and the camber angle. The two angles can be measured separately, however, by using a proprietary alinement gauge of the type just described, and again, in the interests of accuracy when dealing with measurements of only a degree or so, a reliable alinement gauge must be considered almost essential.

On some independent suspension systems the camber angle may be adjusted by rotating the pivot pins of the upper

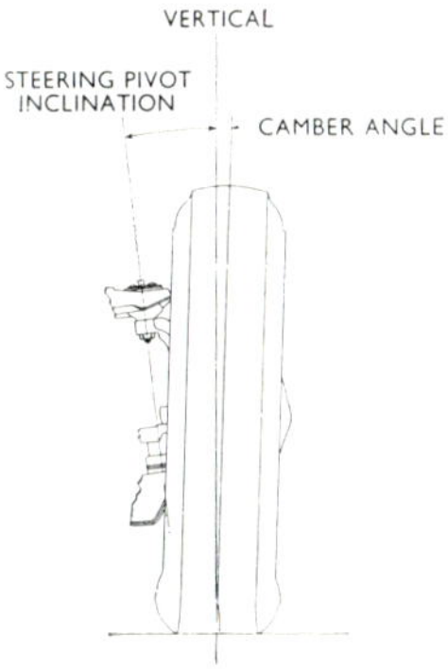

FIG. 86. CAMBER ANGLE AND STEERING PIVOT INCLINATION

wishbones in the linkage, these having eccentric bearing surfaces. If, as is usual, the pivots are threaded, alteration of the camber angle will affect the castor angle to some extent, but it is generally possible to achieve the necessary compromise. When no adjustment is provided, and on conventional front axles, the angle can be altered only by the use of hydraulic bending tools applied to the axle beam or suspension arms.

The Effect of the Ackerman Angle. With this layout, the steering arms attached to the stub axles converge towards one another when in the straight-ahead position. This, on a corner, results in the inner wheel being turned through a greater angle than the outer wheel, thus following a curve of smaller radius, as is necessary if tyre wear is to be avoided and satisfactory steering is to be obtained. If one steering arm is bent as the result of a minor collision the correct relationship of the wheels when in the straight-ahead position can probably be restored by readjusting the length of the track rod : but this will not ensure the correct relationship when cornering. The only cure, therefore, is to remove the steering arm in order to check it in accordance with the manufacturer's drawing or data sheet. Even if the arm should be only slightly bent it must not be straightened ; a replacement must be fitted.

Checking the Front Axle on the Bench. Too much emphasis cannot be placed on the necessity for working to the manufacturer's dimensions when checking the front axle or any part of the stub axles and steering heads. The actual process of checking the various parts is straightforward once the necessary data have been obtained. Special jigs are usually adopted to check the truth of the axle itself. The smaller components are checked by measurement or by templates.

Steering Gearbox Adjustments. The majority of cars nowadays are fitted with steering mechanisms manufactured by specialist firms, exceptions being some types of worm and wheel or worm and sector gears, and rack-and-pinion mechanisms used on some independent front wheel suspension systems. The large measure of standardization, however, has rendered it possible to tabulate in Chart No. 19 the chief causes of backlash in each type of gear, and the appropriate corrective action.

The normal service adjustment provided is sufficient to compensate for the wear which develops during a very large mileage. When the adjustment is no longer effective, or oil leakage or other troubles develop, it is necessary to dismantle the steering gearbox to renew the worn components.

It may be mentioned that full instructions for this work, which is outside the scope of this book, will be found in *Automobile Steering, Braking and Suspension Overhaul*, by the same author. In addition, the checking and overhauling of front axles, swivel pins, springs, independent suspension systems, hubs and similar components which vitally affect the steering are covered fully.

Power-assisted Steering. The growing tendency to provide some form of power assistance in the steering gear of larger cars has resulted in the development of several specialized

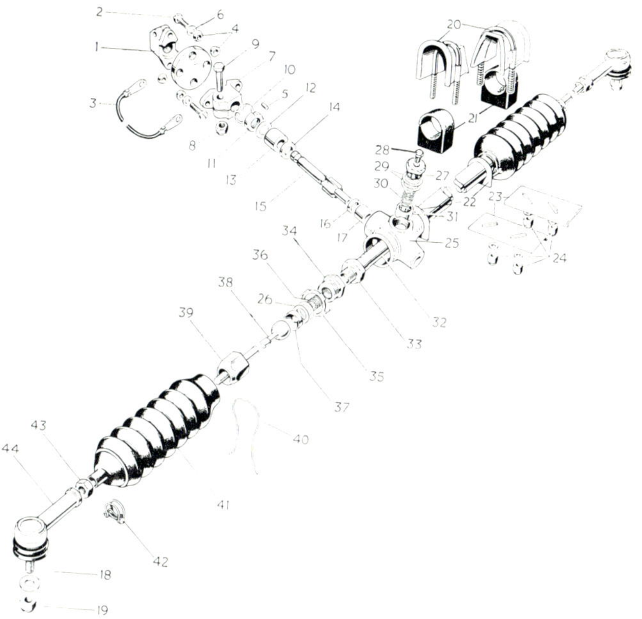

FIG. 87. TYPICAL RACK-AND-PINION STEERING GEAR FAULTS

1. Steering coupling (upper) loose
2. Bolt loose
3. Earthing cable—bad connexion
4. Rubber bushes for coupling worn
5. Dowel loose
6. Washer missing
7. Steering coupling (lower) loose
8. Nyloc nut loose
9. Coupling pinch-bolt loose
10. Circlip incorrectly fitted or broken
11. Pinion retaining ring
12. Adjusting shims incorrectly fitted
13. Pinion shaft bush worn
14. Thrust washer worn
15. Pinion shaft and pinion worn
16. Pinion thrust washer worn
17. Pinion shaft bush worn
18. Washer missing
19. Nyloc nut for tie-rod end loose
20. U-bolts retaining rack housing loose
21. Rubber bushes worn
22. Abutment plates distorted
23. Locating plates damaged
24. Nyloc nuts for U-bolts loose
25. Rack housing damaged
26. Tie-rod joint incorrectly adjusted
27. Damper cap damaged
28. Grease plug missing
29. Damper shims incorrectly adjusted
30. Damper spring weak or broken
31. Damper plunger binding
32. Rack worn
33. Joint lock-nut loose
34. Joint sleeve nut loose
35. Locking tab not locked
36. Joint spring weak or broken
37. Joint cup worn
38. Tie-rod ball joint worn
39. Joint cup nut loose
40. Locking wire missing
41. Rubber gaiter torn
42. Clip loose
43. Lock-nut loose
44. Tie-rod end worn

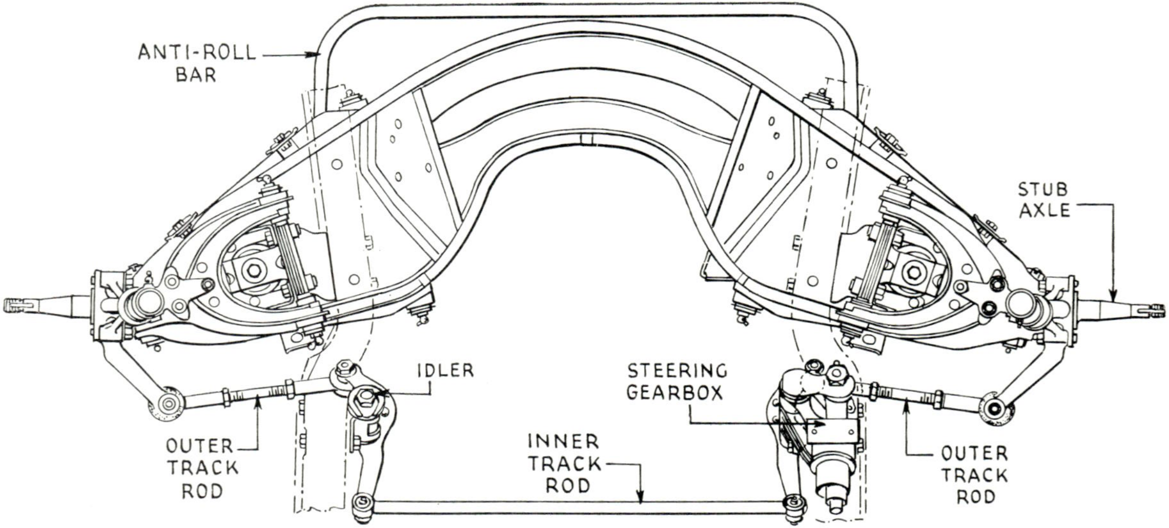

Fig. 88. Typical Modern Steering Linkage, with Divided Track Rods, the Outer Rods being Adjustable

layouts in which oil is supplied under pressure from an engine-driven pump to a servo piston working in a cylinder which may form an integral part of the steering gearbox or which may be

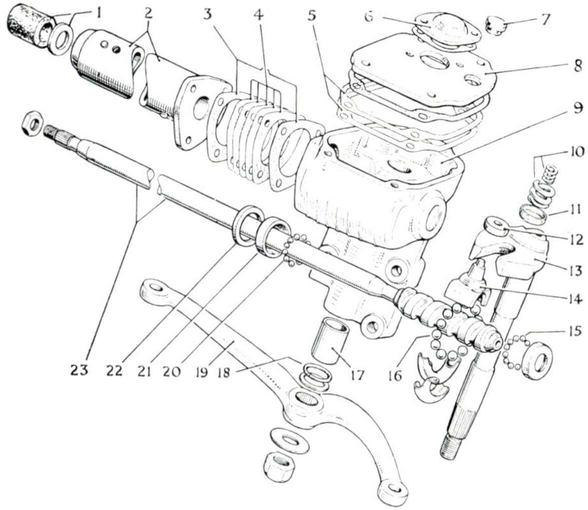

FIG. 89. POSSIBLE FAULTS IN A RECIRCULATING-BALL STEERING GEARBOX

This is a Burman "F" type unit. Other designs are broadly similar

1. Felt bush and washer worn
2. Outer column not securely held
3. Joint washers leaking
4, 5. Incorrect number of shims fitted
6. Spring cap loose
7. Filler plug missing
8. Top cover distorted
9. Steering box damaged
10. Double coil spring weak
11. Damper button incorrectly adjusted
12. Guide roller worn
13. Rocker shaft worn. Splines damaged
14. Nut worn
15, 16. Steel balls worn or incorrectly installed
17. Rocker shaft bush worn
18. Oil seal leaking
19. Drop arm (swing lever) bent or splines worn or damaged
20, 21. Upper bearing worn
22. Distance piece omitted
23. Inner column and worm worn

coupled between the frame and some convenient point in the steering linkage.

When trouble develops in a power steering system, it is obviously desirable to obtain the appropriate fault-tracing and servicing instructions from the manufacturer of the car or of the steering gear. The tests shown in Chart No. 20, however, will enable the more obvious (and most likely) causes of failure to be eliminated.

No attempt should be made to dismantle the pump or the servo components without first consulting the service manual. Fortunately, failure of the power assistance does not normally immobilize the car, the system being so designed that normal manual control will be retained, although the effort required at the steering wheel will, of course, be rather heavier when sharp turns are made at low speeds.

CHAPTER X

BRAKING SYSTEM TROUBLES

THE tendency of manufacturers nowadays to fit proprietary braking systems such as the Lockheed and Girling designs, greatly simplifies servicing and fault-tracing. An additional advantage is that standardized replacements are readily available, including reconditioned, re-lined brake shoes; the correction of any troubles which may arise is, therefore, a simple proposition.

Chart No. 21 is applicable not only to these standardized systems, but also to the various designs of mechanically-operated brake fitted by some individual manufacturers to earlier cars. In the main the chart is self-explanatory, but, as with charts in other chapters, amplification of some points may aid diagnosis and fault-tracing. At the outset it should be mentioned, however, that the manufacturers of standardized braking systems have produced comprehensive instruction leaflets which will repay detailed study if major jobs, such as relining the brakes, overhauling the components of hydraulic systems, or disconnecting, freeing, greasing, reassembling and re-setting the operating linkage of mechanically-operated brakes, are to be undertaken.

Adjusting the Brakes. On many modern braking systems any excessive clearance between the friction linings and the drum or disc is automatically taken up. In such cases, and also when manual adjustment of drum-type brakes is necessary, running adjustments are carried out at each brake drum, obviating any need for altering the settings of operating rods, compensators or other parts of the linkage; in fact, these parts should be re-set only when a major overhaul of the braking system is undertaken, at fairly lengthy intervals.

The methods of manual adjustment fall into two broad groups: adjusters in the form of hexagonal nuts or squared spindles projecting from the brake backplate, the spindle sometimes being fitted with a small tommy-bar or handwheel for convenience in turning it; and notched disks attached to an

CHART No. 21

BRAKING SYSTEM TROUBLES

- **Loss of Efficiency** — First check brake adjustment as described in text
  - Hydraulically-operated Brakes
    - Pedal travel excessive: becomes normal when pedal is " pumped " several times
      - Brake shoes require adjustment.
      - Wear at one or more joints.
      - Master cylinder cup worn.
    - Pedal feels " springy "
      - System requires " bleeding " (see text).
      - New linings not bedded in.
      - Poor quality linings.
      - Master cylinder attachment loose.
      - Brake drums weak.
    - Normal pedal action
      - Greasy linings (see below).
      - Linings worn out.
      - Poor quality linings.
      - Badly scored or distorted drums or discs.
      - Shoes incorrectly centred in drums.
  - Mechanically-operated Brakes — Assuming pedal travel is not excessive
    - Brake levers or operating mechanism incorrectly set (see text).
    - Binding cables or expanders.
    - Shoes incorrectly centred in drums.
    - Brakes not balanced (see below).
    - Greasy linings (see below).
    - Poor quality linings.
    - Linings badly fitted.
    - Distorted or badly scored drums.
    - New linings not yet bedded in.

- **Loss of Efficiency—brake mechanism and drums in good condition** — Check for oil or grease in drums or on discs due to over-lubrication of rear axle, hubs or brake gear, or to faulty oil seals (see Chart No. 16) — Water in brakes after washing car or driving through water-splash. Apply brakes while driving to heat up drums and evaporate water
  - Linings oily or greasy.
  - Excessive play on wheel bearings (Chart No. 16).
  - Excessive heating of brakes due to prolonged use during steep gradients or fast driving (" brake fade ").
  - Defective servo, if fitted (Chart No. 22).

- **Unbalanced Braking or Snatch** — Check adjustment at each wheel. Verify by road test (see text)
  - Unequal tyre pressures.
  - Unequal brake adjustment.
  - Compensator gear not effective.
  - Uneven wear of linings (see below).
  - Oil, grease or water in one or more drums (see above and Chart No. 16).
  - Dust and grit in drums.
  - Oval or scored drums.
  - Distorted or scored discs.
  - Road spring fractured or shackles worn.
  - Brake backplate loose on axle.
  - Leading edge of lining picking up—requires chamfering.
  - Different types of lining on various shoes.

Brakes Binding
- Check adjustment of brakes and operation of servo, if fitted (Chart No. 22)
  - Hydraulically-operated Brakes
    - Blocked or partly obstructed pipe lines.
    - Vent hole in reservoir filler cap blocked.
    - Piston cup in master cylinder or wheel cylinder swollen.
    - Mechanical faults (see below as for mechanically-operated brakes).
    - Incorrect adjustment of push-rod connecting pedal to master cylinder.
  - Mechanically-operated Brakes
    - Check operating cables or rods for freedom of movement
      - Weak or broken brake shoe pull-off springs.
      - Shoes seized on anchor pins.
      - Pedal return spring weak, disconnected or broken.

Brakes Overheating or Smoking
- Check adjustment and make sure shoes are not binding
  - Most likely cause is brakes binding (see above).
  - Prolonged use of brakes during steep descent or fast driving.
  - Brake drums oval.
  - Brake drum diameter too great due to excessive grinding (see " Rapid or unequal wear of linings " below).

Brake Judder or Squeal
- The two faults are closely associated, judder being caused by low-frequency vibrations and squeal or screech by high-frequency vibrations
  - Grease or oil on linings.
  - Leading edge of lining picking up.
  - Linings badly fitted.
  - Brake shoes out of line.
  - Rivets in contact with drum.
  - Worn shoe pivot anchorages.
  - Weak drum—may be due to excessive grinding.
  - Loose brake backplate.
  - Silencing band—if fitted—omitted from brake drum.
  - Damper weights—if fitted—omitted from shoes.

Rapid or Unequal Wear of Linings
- Rapid, but approximately equal wear on all linings
  - Brakes binding.
  - Dust and grit in drums.
  - Brake drums or discs scored.
  - Unsuitable linings.
  - Violent braking.
- Rapid wear of one or more linings in individual drums
  - One or more brakes incorrectly adjusted or binding.
  - Oil or grease in one or more drums or on discs throws excessive load on clean linings.
  - Worn wheel bearings.
  - Excessive grinding of drums results in contact only at centre of lining.
  - Different grades of lining on individual shoes.

expanding sleeve which separates the brake shoes, or to the end-caps of the wheel cylinders on hydraulic brakes, access to the notches on the adjuster being obtained through slots in the brake drum or in the brake backplate.

Spindle and Handwheel Adjustments. In the first group, the road wheel is first jacked up, and the adjuster screwed inwards in a clockwise direction until the brake shoes touch the drum. The adjuster is then unscrewed until the drum is just free,

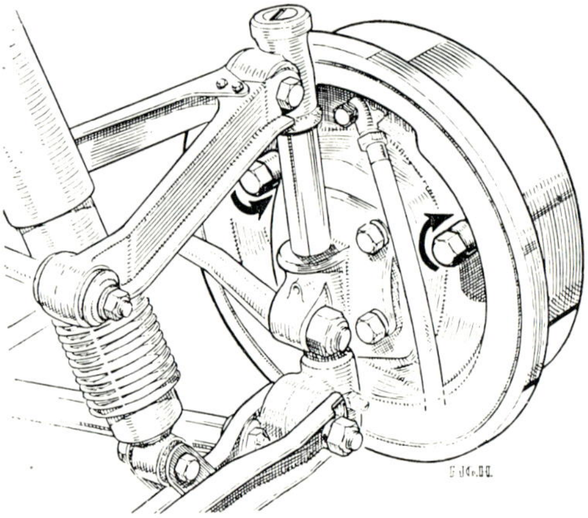

FIG. 90. WIDELY-USED FRONT-BRAKE ADJUSTMENT

but the shoes do not bind. On Lockheed brakes fitted with two hexagonal adjusters, these must be rotated in opposite directions, as shown in Fig. 90.

On some brakes the correct clearance between the shoes and the drum is automatically obtained by unscrewing the adjuster a specified number of "clicks," which are heard, as the shoe-expanding plungers drop on to flats on the adjuster. This applies to Girling brakes, on which the adjuster should be unscrewed by one click, or by two or three clicks if the brake still binds; on Girling Hydro-mechanical Two-leading-shoe brakes and Cowdrey brakes the adjuster should be slackened off to the extent of five clicks.

On the types of Lockheed brake provided with a small

handwheel adjuster the handwheel should be turned as far as possible in a clockwise direction and then released—not turned back—the correct clearance being automatically obtained. A finer adjustment can be obtained if an assistant depresses the foot brake pedal while the adjustment is made.

Strictly speaking, it is unnecessary to jack up the wheel when an adjuster is fitted which automatically provides the

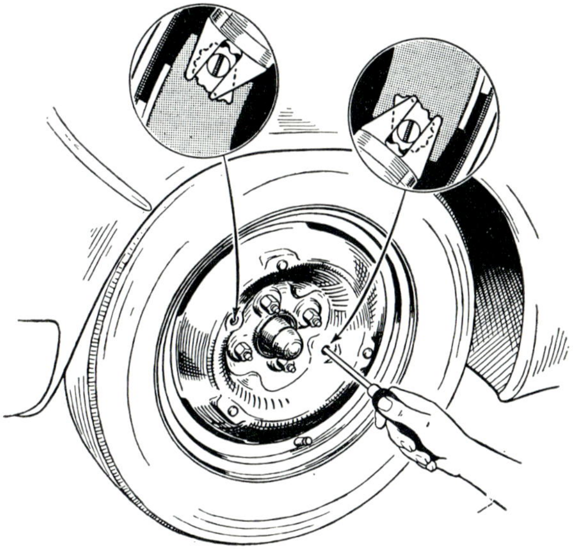

FIG. 91. METHOD OF ADJUSTMENT OFTEN FITTED TO LOCKHEED BRAKES

correct clearance, but this is always advisable in order to make sure that the shoes are not binding owing to one or more of the defects mentioned under " Brakes Binding " in Chart No. 21. An occasional check is also advisable when an automatic adjustment for wear is incorporated, as on Girling Hydro-mechanical brakes (though not the Two-leading-shoe design).

Notched Disk Adjusters. When a notched disk is attached to a threaded shoe expanding sleeve or to threaded caps on the end-caps of the brake expanding cylinder, access to the disk may be obtained either through a slot in the brake backplate, after removing a dust-excluding cover, or through a notch in the wheel, after removing the nave plate and rotating the wheel to bring the slot into line with the adjuster.

Sometimes a slot is provided both in the wheel and in the backplate. While the slot in the wheel is more accessible, that in the backplate must be used when carrying out an adjustment while the wheel is rotating, and also when certain types of brake testing equipment are used.

In both instances a screwdriver is inserted through the slot and the notched disk rotated in the appropriate direction to bring the shoes into contact with the drum, after which the adjustment should be slackened off until the drum is just free to rotate.

Centralizing Shoes in Drums. On some types of brake it is essential that the expander or the adjuster housing should be free to " float " slightly in the backplate in order to allow the brake shoes to take up the correct position in the drum. On Girling brakes (except Two-leading-shoe types) and on Lockheed external-cylinder brakes, it is the expander which should be free to move ; the retaining nuts should not, therefore, be fully tightened. On Lockheed brakes provided with a handwheel adjustment the adjuster housing should be free to float. Mud and grit should be cleared away and a penetrating oil applied before carrying out the adjustment. On Girling Two-leading-shoe brakes it is advisable to slacken off the nuts which secure the adjuster unit and to screw up the adjuster until the shoes are tightly expanded in the drum. Then tighten the adjuster securing nuts firmly before slackening off the adjuster by two clicks.

A further point is that on some Bendix systems the shoes must be correctly centralized by measuring the clearance between the shoe and the drum by passing a feeler gauge through a slot in the backplate. This adjustment is required only when re-lined shoes are fitted. The correct clearance should be 0·008 in., at a point approximately one inch from the end of the shoe lining and it is obtained by slackening the anchor pin nut and tapping it in the required direction.

Oval and Scored Drums. It often happens that when adjustment is carefully carried out, and the brake shoes are centralized, the drum can be heard to touch the shoes momentarily at regular intervals when the wheel is rotated. Frequently it is necessary to slacken off the adjustment to an extent which seriously reduces braking efficiency before the drum rotates freely. This is a symptom of a distorted, or badly scored

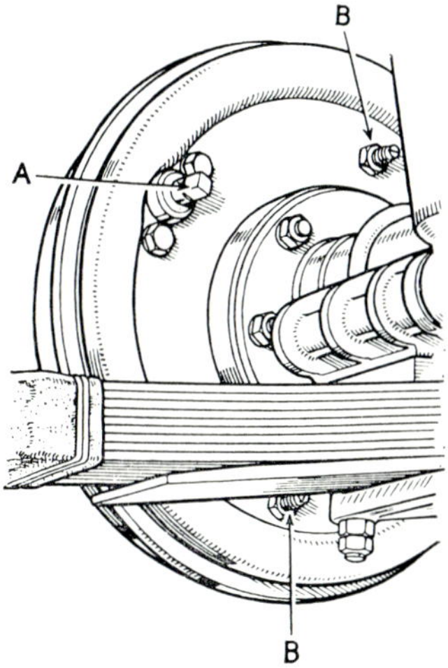

FIG. 92. COMMON FORM OF REAR BRAKE ADJUSTMENT

A, square-ended spindle of adjuster. *B*, shoe steady posts (not fitted to all brakes). Do *not* adjust these except when fitting replacement shoes

drum. The only remedy is to fit a new drum or to have the existing drum reground to restore a smooth surface.

If it is necessary to remove an appreciable amount of metal to restore circularity or to remove deep scores, however, the drum may be so weakened that it will quickly distort again under the stress of heavy braking. An additional disadvantage is that the increased diameter of the drum will result in the brake linings making contact with it only over their central portions, leading to poor braking and rapid wear.

An excellent alternative to fitting new drums is to have the existing drums ground out to take hardened steel liners, provided that sufficient thickness of metal exists ; hardened liners not only have a very long life, but at the same time provide greater braking efficiency than is obtainable with standard drums.

Brake Testing. Although the brakes may be carefully

FIG. 93. BRAKE DRUM REMOVED FOR INSPECTION

Look for accumulation of brake lining dust, scored contact surface and ovality of the drum

adjusted as just described, it may be found, after a road test, that some of the brakes are more powerful than the remainder, resulting in a wheel locking and a tendency to swerve when the brakes are applied. A test should be made, therefore, on a straight stretch of road with a good surface, the brakes being applied with various degrees of violence at ten, twenty,

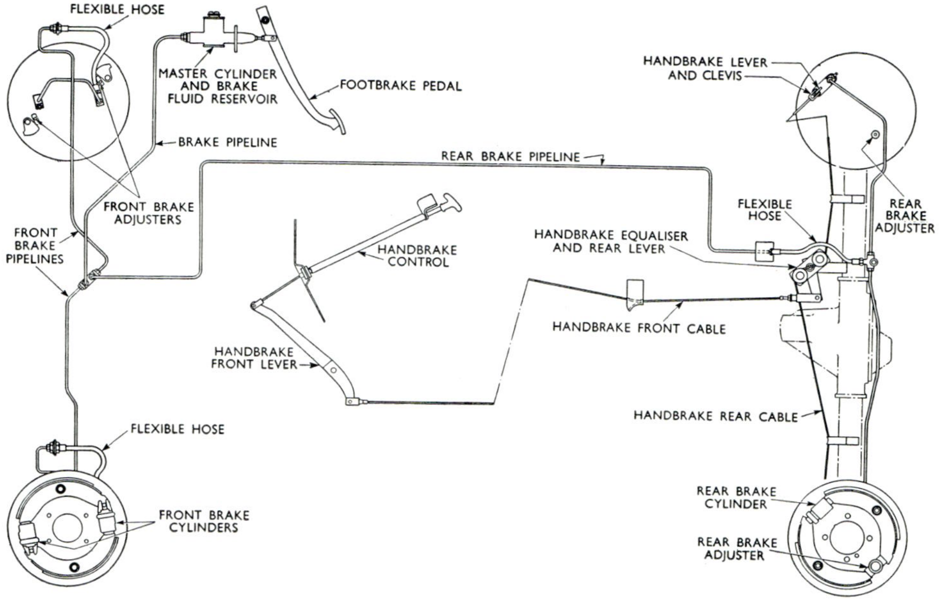

Fig. 94. Layout of a Typical Hydraulic Braking System

thirty and forty miles per hour. Tests may also be advisable at higher speeds if the performance of the car warrants it. The cause of any tendency to swerve, or of a wheel to lock, or a brake to overheat, should be diagnosed by referring to the appropriate section of Chart No. 21.

The actual braking efficiency can be measured during the test with a pendulum-type brake efficiency meter, such as the well-known Ferodo-Tapley design, which records braking power both in terms of percentage efficiency and as the distance required to pull up from a speed of 20 m.p.h. The relationship between braking efficiency and stopping distance is as follows :

Distance to Stop From 20 m.p.h.	Efficiency (per cent)	Practical Assessment
45 ft	30	Dangerous
34 ,,	40	Fair
27 ,,	50	Good
22 ,,	60	Very good
19 ,,	70	Excellent

An efficiency of 70 per cent simply means that the retarding force exerted by the four wheels is equal to 70 per cent of the total weight of the car. Efficiencies as high as 100 per cent have been recorded under " crash stop " conditions on cars on which the brakes were in perfect condition, and the tyres possessed an excellent tread, but such efficiencies are seldom met with, or needed, under normal driving conditions.

Most types of garage brake testers are calibrated in terms of percentage efficiency, although some measure the retarding force in pounds " pull " exerted by each wheel in resisting rotation by power-driven rollers or an endless belt on which the wheel rests during the test. Since any form of garage tester, whether of the " static " type just described, or the " dynamic " type on to which the car is driven and braked, must necessarily make the test at a relatively low speed, a road test is always advisable in addition.

Air in Hydraulic Braking System. Among likely faults in hydraulically-operated braking systems, one of the most common is entry of air into the pipelines. This can occur if the fluid level in the supply tank is allowed to fall too low, or if for any reason a pipeline is disconnected. The air, being compressible, gives a " spongy " action to the brake pedal.

The condition is cured by attaching a length of rubber tubing to a nipple on the brake backplate and "bleeding" the air from the system by allowing fluid to flow from the tube into a clean glass jar. The nipple is usually protected by a rubber cover. On Lockheed systems it may be necessary to remove a sealing plug and to screw a special nipple in place.

The bleeder valve is then opened by about one turn—a special

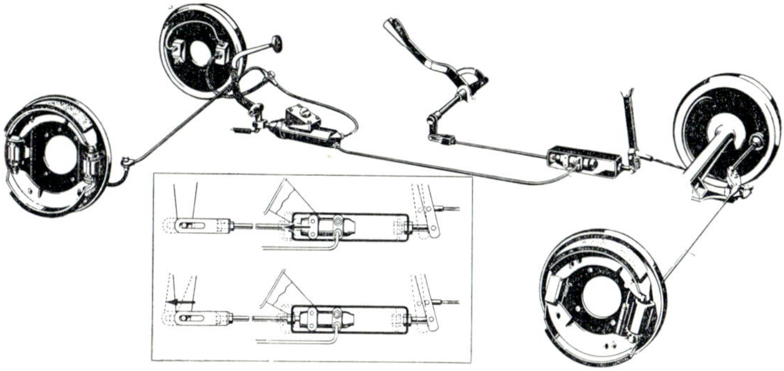

FIG. 95. LOCKHEED HYDRAULIC-MECHANICAL BRAKING SYSTEM USED ON THE AUSTIN A30 AND A35

The rear brakes are actuated by pull-rods connected to the stirrup that surrounds the operating cylinder (*shown inset*). The handbrake also acts on this stirrup

spanner may be required—and the brake pedal is depressed quickly and allowed to return slowly. Pumping should be continued until the fluid issuing from the tube is free from air-bubbles. It will, of course, be necessary to top up the supply tank at intervals to compensate for the fluid which is pumped out of the system.

When bleeding Lockheed and some Girling systems, the brake pedal should be depressed quickly and allowed to return slowly as just described. When bleeding Girling CV master cylinders, which can be recognized by the aluminium body instead of the cast-iron body used on other Girling cylinders, however, the pedal should be pushed down through a full stroke, followed by three short rapid strokes, after which the pedal should be allowed to return quickly to its stop with the foot completely removed.

When a skid-control valve is incorporated in the hydraulic

CHART No. 22

BRAKE SERVO MECHANISM DEFECTS

The faults listed relate only to the power-servo components. General braking system faults (Chart No. 21) may be contributing factors.

- **INSUFICIENT SERVO ASSISTANCE**
  - **VACUUM SERVO SYSTEMS** — Test action of servo by depressing brake pedal and starting engine: marked decrease in foot pressure required on pedal should be evident. If servo is inoperative or insufficiently powerful:
    - Test for air leakage into pipe-line or servo cylinder (see text)
      - Air leak present
        - Loose or damaged unions in vacuum pipe.
        - Air leak at servo cylinder and cover.
        - Air leak past defective piston leather.
      - No air leak detected
        - Collapsed vacuum hose at connexion to engine manifold.
        - Carbon on vacuum valve seating.
        - Incorrect adjustment of valve-operating linkage.
        - Faulty hydraulic master or slave cylinder.
    - When servo is supplied from vacuum reservoir, check reservoir and pipe lines (see text)
      - No loss of vacuum
        - (as for "No air leak detected" above)
      - Loss of vacuum
        - Loose or damaged union in pipe line from induction manifold to reservoir or reservoir to servo.
        - Non-return valve in reservoir leaking.
        - Air leaks at joints of reservoir.
  - **HYDRAULIC PUMP SERVO SYSTEMS** — Check hydraulic fluid system for faults; if sound, fault probably lies in pump
    - Fluid system faults
      - Lack of fluid in supply tank.
      - Loose connexions.
      - Faulty servo valve in master cylinder.
      - Insufficient air pressure in hydraulic accumulator.
      - Faulty cut-out valve.
    - Hydraulic pump faults: testing and overhaul calls for specialized equipment
      - Pump drive from engine or transmission faulty.
      - Defective valves in pump.
      - Leakage at gland.
      - Wear of cylinders, plungers, driving shaft and bearings, shoes, driving pins, etc.

system to limit the pressure applied to the rear brake cylinders, it is necessary first to bleed the front brakes, using the normal pedal action and then to apply the handbrake firmly and to bleed the rear brakes by depressing the pedal slowly through a full stroke and allowing it to return slowly, with a pause of 3–4 seconds between each stroke. This should prevent the ball valve in the restrictor in the rear brake pipe closing, so that

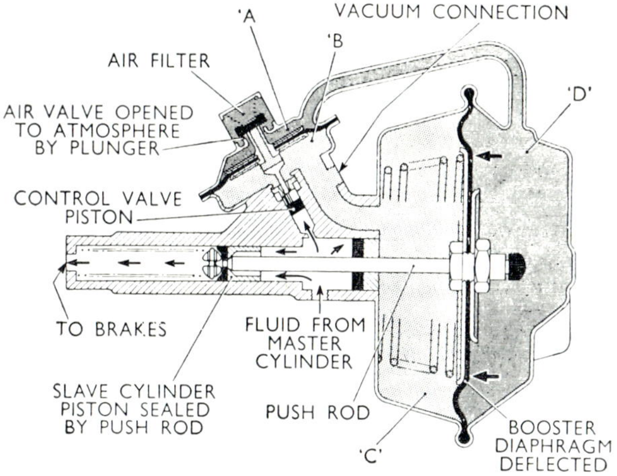

FIG. 96. A MODERN, COMPACT TYPE OF VACUUM SERVO

The air valve has opened, admitting atmospheric pressure to chambers "A" and "D." Chambers "B" and "C" are subject to the partial vacuum in the inlet manifold of the engine. The booster diaphragm is therefore deflected and the push-rod moves the slave cylinder piston to raise the pressure in the brake pipelines

fluid can flow through the valve. If the valve does close, which can be easily detected, since it will not be possible to push the brake pedal completely to the floor, the bleed screw must be tightened without delay while the assistant who is operating the brake pedal releases the handbrake. The combination of these actions will displace the ball valve from its seat. The handbrake must then be reapplied and bleeding continued.

In garages and service stations a special filler is often attached to the reservoir to maintain the fluid level automatically. An alternative method of bleeding is to use a

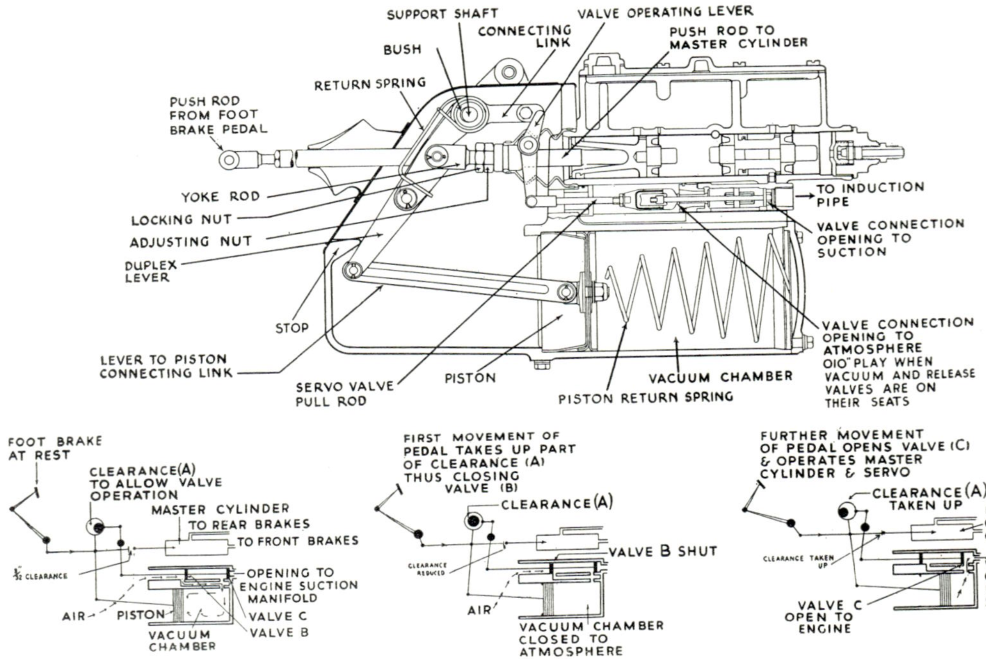

FIG. 97. THE LAY-OUT AND OPERATION OF A VACUUM-SERVO UNIT COUPLED TO THE MASTER CYLINDER OF A HYDRAULIC BRAKING SYSTEM

pressure-tank which is connected to the master cylinder by means of a hose and adaptor. Fluid is thus forced into the system under pressure, obviating the need for pumping the brake pedal and considerably speeding up the operation.

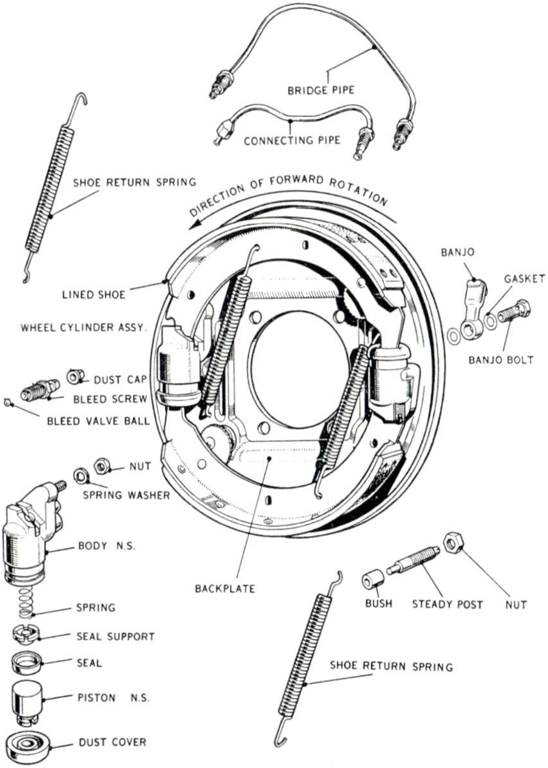

FIG. 98. GIRLING TWO-LEADING-SHOE HYDRAULIC BRAKE

Unsuitable Fluid and Choked Hoses. If the fluid which is drawn off during the bleeding operation is discoloured, dirty, or appears viscous or gummy, each brake should be "bled" until fresh fluid has been pumped throughout the system. The use of mineral oils and other unsuitable fluids is the most likely cause of choked pipelines and swelling of the rubber cups in the master cylinder and wheel cylinders. This results in binding brakes, since the low pressure exerted by the brake-shoe return springs is insufficient to force the fluid back

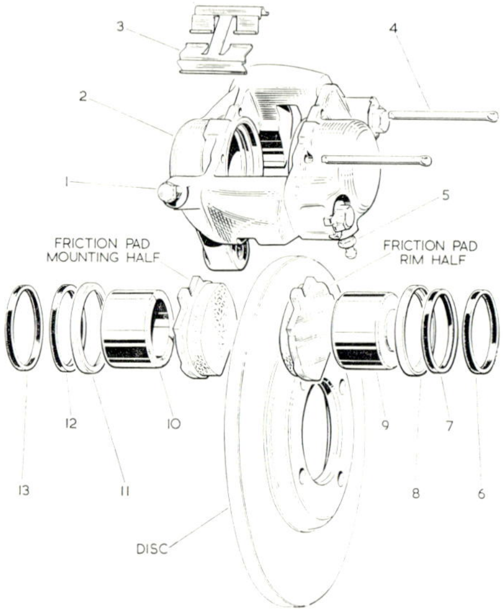

FIG. 99. LOCKHEED LIGHT-DUTY DISC BRAKE

The following are possible faults in addition to worn friction pad linings and a scored or distorted disc—

1. Plug loose
2. Caliper sub-assembly damaged or wrongly assembled (the two parts should not be dismantled in normal service)
3. Pad retaining spring broken or rusty
4. Split pins rusty
5. Bleeder screw not tightened down
6. Fluid seals leaking
7. Dust seals damaged
8. Dust seal retainer damaged
9. Piston rusty and sticking in bore

through the hoses to the master cylinder when the foot pedal is released, or to move swollen wheel cylinder or master cylinder cups which have jammed in the cylinders. The only cure is to dismantle the parts and to fit new rubber cups or flexible hoses, as the case may be; this is not a job to be tackled without a certain amount of experience.

Re-lining the Brakes. Mention has already been made of the fact that the manufacturers of proprietary braking systems supply re-lined shoes to which the linings are riveted by machine tools and are accurately ground to the correct radius. Re-lining thus becomes simply a matter of removing the old shoes and fitting the replacements. There is little justification nowadays for cutting linings to length, drilling them and riveting them to the shoes, unless a well-equipped workshop is available ; badly fitted linings can be the cause of the majority of braking troubles, as will be seen from Chart No. 21.

If it is essential to fit new linings to existing shoes, however, the die-pressed type, generally supplied in sets, pressed to the correct radius, and with the rivet holes accurately drilled, should be used whenever possible. A suitable clamp to hold the lining firmly to the shoe while riveting, and a rivet set and anvil of the right size, are essential.

Disc Brakes. When disc brakes are fitted, fault-tracing is basically similar to the principles already outlined for normal hydraulic braking systems. The chief difference, however, is the ease with which the brake lining segments or pads can be removed from most types of disc brake for inspection or renewal. Moreover, adjustment for lining wear is automatic and manual adjustment is normally required only to take up any stretch that may develop in the hand-brake cables, assuming that disc brakes are fitted to the rear in addition to the front hubs.

CHAPTER XI

SUSPENSION FAULTS

WHEN diagnosing suspension faults it must be borne in mind that the springs and shock absorbers, or shock dampers, as they are often termed nowadays, are complementary. Often suspension troubles can be attributed to the shock absorbers rather than to the springs, while lack of effective damping of spring movements can also cause broken spring leaves.

A further point is that spring and shock absorber faults can also affect the steering and braking performance of the car, so that cross reference between Charts Nos. 18 to 24 may be necessary to determine which of a number of possible faults is responsible for a particular trouble.

Leaf Springs. When laminated leaf springs are fitted, they may or may not require regular lubrication. It is essential to establish this point, either from the instruction book or by consulting the local agents for the car, as modern springs usually have rubber or plastic wear-resistant pads at the tips of each leaf, or interleaving strips between the leaves. Either arrangement renders lubrication unnecessary—and even harmful, if rubber is used, since this will quickly deteriorate in contact with oil or grease. Also, the design of the suspension often takes into account the self-damping properties of a leaf spring. If the spring is rendered too supple by lubrication of the leaves, the roadholding of the car may be affected and there may be a risk of the spring leaves being broken by excessive deflection.

In some cases, on the other hand, a manufacturer will specify regular lubrication of the springs with oil or grease and if this is neglected, the springs may become excessively stiff, causing harsh suspension.

Recognizing Spring Defects. Regular cleaning and inspection of the springs is a wise policy, since any defects will be detected before major trouble develops. If one spring becomes more flattened than the remainder, or (if they are flat springs or springs having a reverse camber), deflects to a greater extent, it is advisable to scrub the edges clean and to examine them for signs of a broken leaf. It is worth remembering, however, that on some cars one leaf is split, and shows a pronounced gap on either side of the centre bolt at a short distance away from the bolt.

The modern tendency to fit flat springs, or springs which have a reverse or upward camber, renders it less easy to detect the normal loss of resilience which occurs after a considerable mileage. Even the position of the swinging shackle is no longer a reliable guide, although it should usually be vertical when the car is carrying its designed load. The end attached to the spring moves forwards or rearwards, so that the shackle is inclined from the vertical as the spring weakens, the direction depending on whether the spring is of the flat or reversed cambered type, or should possess a normal camber. The safest plan, however, is to ascertain the correct camber from the manufacturer or his local agent.

Worn Shackle Pins and Bushes. The detection of worn shackle pins and bushes is a straightforward matter when these are of the plain-bushed type, but a cause of rapid shackle pin and bush wear which is often overlooked is fitting the shackle pins with the oil feed holes in the wrong position. The hole must always face the side of the bush which receives the least load : the holes in the pins fitted to the swinging shackle, for instance, should face one another, that in the upper pin pointing downwards, and the hole in the lower pin pointing upwards. On a number of cars, however, threaded shackle pins and bushes are used ; these wear much less rapidly, while the threads also provide positive sideways location of the shackle.

On plain-bushed shackles the side-clearance of the shackles is often controlled by thrust washers. Excessive side clearance can cause a pronounced knock, so that it is essential to make sure that it is within the figure recommended by the manufacturer. It should seldom exceed 0·015–0·025 in., while between 0·003 in. and 0·005 in. generally represents the minimum permissible clearance. In some cases thin pen-steel washers can be obtained which allow the clearance to be accurately adjusted ; otherwise plain steel washers must be carefully filed or rubbed down to give the correct clearance.

Yet another arrangement is a distance-tube fitted over the shackle pin and gripped between washers fitted between its

CHART No. 23 DIAGNOSING SUSPENSION TROUBLES

- **Harsh Suspension—Excessive Road Shocks Transmitted to Car**
  - **Normal Suspension Systems** — Check condition of springs—if rusty spray with penetrating oil (but see Over-Flexible Suspension, below)
    - If spring action is not improved
      - Spring leaves corroded, causing excessive friction.
      - Tips of leaves have worn depressions in leaves above.
      - Lubricating pads or steel rollers at tips of leaves worn out.
      - Lubricant in gaiters has hardened when gaiters are fitted, preventing fresh lubricant from reaching spring.
      - Extra-load or rebound leaves fitted wrong way up, increasing stiffness of spring when normally loaded.
      - Shackle pins seized.
      - Shock absorber defects (Chart No. 24).
  - **Independent Suspension Systems**
    - Leaf spring systems
      - Faults as for normal suspension above, except shackle pin seizure.
      - Suspension arm pivots seized.
    - Coil spring systems
      - Coil spring tension too great—but usually revealed chassis by being too high from ground.
      - Suspension arm pivots seized.
      - Shock absorber defects (Chart No. 24).
    - Torsion bar systems
      - Torsion bars incorrectly adjusted (see text).
      - Suspension arm pivots seized.
      - Shock absorber defects (Chart No. 24).
  - **Hydro-pneumatic Systems**
    - Pressure in system too high.
    - Faulty valves in suspension units.
  - **Normal Suspension Systems** — Check appearance of springs for signs of flattening, broken leaves, etc. (see text). If springs should not be lubricated, note whether they have been sprayed in error
    - Lack of friction due to lubricatioln of springs where manufacturer specified that eaves should not be lubricated.
    - Weakened springs—loss of camber.
    - Broken leaves.
    - Shock absorber defects.

- **Over-Flexible Suspension**
  - Independent Suspension Systems
    - Leaf spring systems
      - As for normal suspension systems.
    - Coil spring systems
      - Coil spring tension insufficient due to weakened or incorrectly adjusted spring.
      - Shock absorber defects.
    - Torsion bar systems
      - Weakened torsion bar.
      - Shock absorber defects.
  - Hydro-pneumatic Systems
    - Pressure in system too low.
    - Faulty valves in suspension units.
- **Car Sags to One Side**
  - Normal Suspension Systems
    - Weak spring.
    - Broken spring leaves.
  - Independent Suspension Systems
    - Incorrect adjustment of coil springs or torsion bars.
    - Weakened springs or broken leaves in transverse leaf spring systems.
  - Hydro-pneumatic Systems
    - Defective suspension unit(s).
    - Loss of pressure in units on low side of vehicle.
- **Breakage of Springs**
  - Leaf Springs
    - Overloading.
    - Violent braking or grabbing brakes.
    - Incorrect adjustment of torque reaction arms of cables, if fitted.
    - Loose U-bolts.
    - Loose spring clips.
    - Head of centre bolt or dowel does not fit into recess in axle.
    - Broken rebound check straps.
    - Seized shackles.
    - Incorrect heat-treatment of reconditioned springs.
    - Shock absorber defects (Chart No. 24).
  - Coil Springs and Torsion Bars
    - Overloading.
    - Surface of spring or bar damaged or scratched, (even slight corrosion of enclosed torsion bars).
    - Shock absorber defects (Chart No. 24).
- **Noises**
  - Leaf Springs
    - Lack of lubrication.
    - Loose U-bolts.
    - Worn shackle pins and bushes.
    - Side-play on shackles.
    - Broken check straps.
    - Shock absorber defects (Chart No. 24).
  - Coil Spring and Torsion Bar Systems
    - Lack of lubrication.
    - Worn suspension arm bushes or bearings.
    - Worn torsion bar splines.
    - Shock absorber defects.

ends and the spring mounting. If the clearance between the spring eye and the washers is excessive, the tube must be shortened by filing the required amount from one end.

Rapid failure of rubber bushes in the spring eyes and shackles

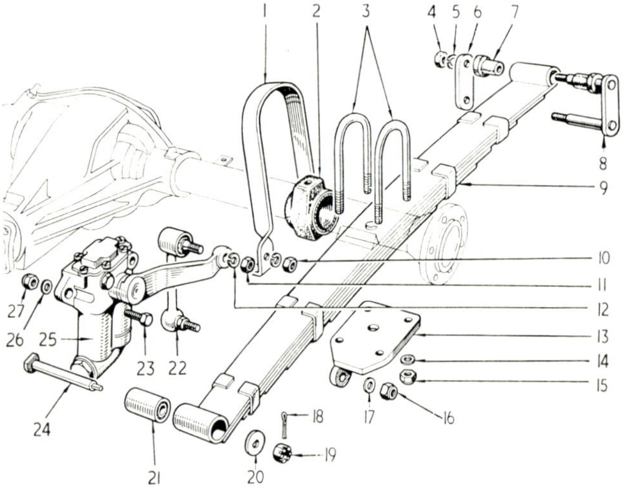

FIG. 100. POSSIBLE FAULTS IN LEAF-SPRING SUSPENSION

1. Bump strap loose or broken
2. Bump and rebound rubber worn
3. U-bolts loose
4. Shackle nut loose
5. Spring washer missing
6. Shackle plate bent
7. Rubber bush worn
8. Shackle pin loose
9. Spring leaves weak or broken
10, 11. Nuts loose
12. Spring washer missing
13. Spring plate distorted
14. Washer missing
15. 16. Nyloc nuts loose
17. Washer missing
18. Split pin missing
19. Slotted nut loose
20. Washer missing
21. Bush worn
22. Damper link bush worn
23. Damper securing bolts loose
24. Shackle pin worn
25. Damper worn or requires topping-up
26. Washer missing
27. Nyloc nut loose

is usually caused either by allowing them to become saturated with oil or grease, or by incorrect assembly when the springs are fitted to the car. The shackle pin nuts must not be fully tightened until the full weight of the car is carried by the springs, and the car has been bounced up and down once or twice to allow the bushes to assume a neutral position in which

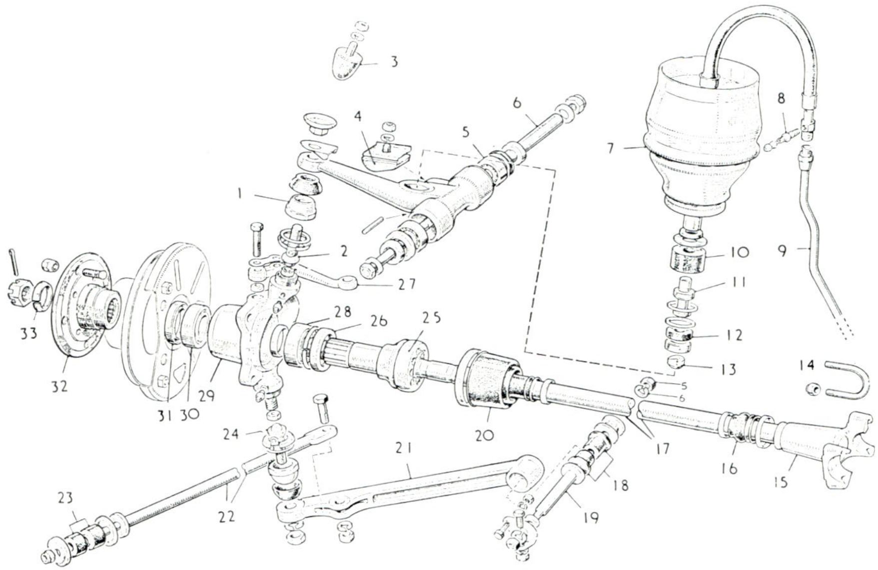

FIG. 101. B.M.C. HYDROLASTIC FRONT SUSPENSION

The suspension shown is fitted to B.M.C. 1100 models. A similar arrangement is used on the Mini and 1800 ranges

1. Ball pin retainer
2. Upper ball pin
3. Bump rubber
4. Rebound rubber
5. Pivot shaft bearing
6. Pivot tube
7. Hydrolastic displacer unit
8. Servicing valve
9. Interconnecting pipe
10. Roller joint seal
11. Foot roller joint
12. Retaining pad
13. Roller joint seat
14. U-bolt
15. Universal joint flange
16. Seal
17. Drive shaft
18. Lower arm bushes
19. Lower arm pivot pin
20. Rubber boot
21. Lower support arm
22. Tie-rod
23. Rubber pads
24. Lower ball pin
25. Constant-velocity universal joint
26. Inner oil seal
27. Steering arm
28. Inner bearing
29. Swivel hub
30. Outer bearing
31. Outer oil seal
32. Driving flange
33. Drive shaft collar

the rubber is free from torsion. Otherwise the bushes will be unduly deflected and the rubber will have a short life.

Checking Independent Suspension Systems

A similar precaution should be adopted when rubber bushes are fitted to the swinging arms or " wishbones " of independent suspension systems. When they are fitted, it may also be

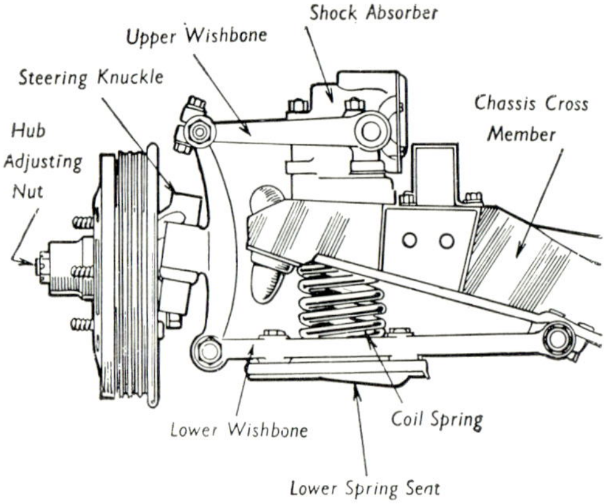

Fig. 102. Typical Parallelogram Independent Suspension System, Making Use of a Coil Spring

necessary to ensure that the front of the car is a definite height from the ground, the tension of the coil spring or torsion bar being adjusted to raise or lower the front end of the chassis as required. On torsion bar suspension systems this adjustment is effected at the fixed end of the torsion bar ; that is, the end remote from the wheel carrier arm.

Hydro-pneumatic Suspension Systems. Fault-tracing on hydro-pneumatic suspension systems, whether of the advanced self-levelling type fitted to Citroën, Mercedes and Rolls Royce models or the simpler B.M.C. Hydrolastic type, should be left to a properly equipped service station. Special equipment is needed to pressurize the suspension units and a thorough working knowledge of the system is essential in order to diagnose and rectify faults.

Shock Absorber Defects

The type of shock absorber fitted depends on the design and characteristics of the suspension system. On most modern chassis hydraulic shock absorbers are employed, their function being, of course, to damp out the natural vibrations or oscillations that are set up in the springs when the wheels strike irregularities in the road surface.

When an earlier model shows symptoms of poor roadholding, pitching on rough surfaces and a generally "lively" suspension,

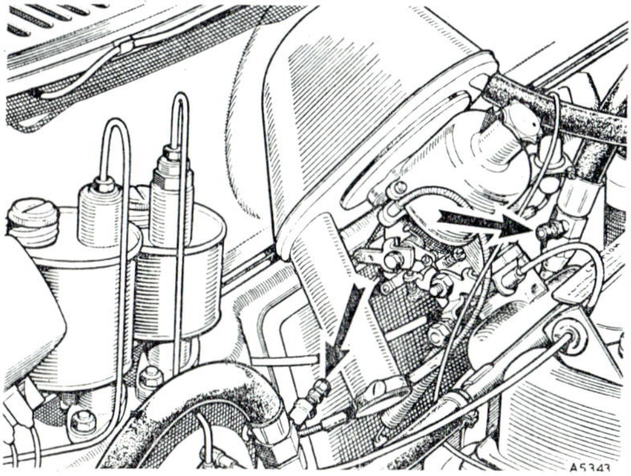

Fig. 103. The Valves shown Should Not be Slackened except when Servicing a Hydrolastic Suspension System

On some models the valves are mounted on the rear sub-frame of the vehicle

it is logical to check the fluid level in the shock absorbers. In many earlier instruction books, in fact, inspection of the fluid level was called for at intervals of between 5,000 miles and 15,000 miles; but nowadays this attention is seldom required and many modern shock absorbers have no provision for topping-up.

In diagnosing faults in modern suspension systems, therefore, it is sufficient to examine the shock absorbers for any sign of leakage and to renew them if there is any evidence that the seals have become excessively worn. In all probability the internal components, including the piston and check valves, will also be worn to such an extent as to render the

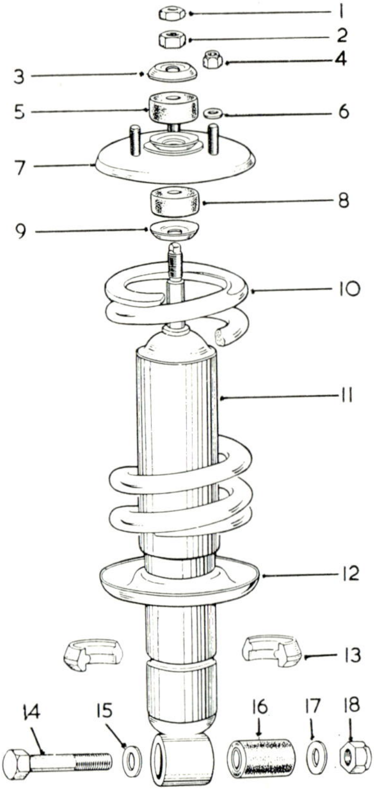

FIG. 104. TELESCOPIC DAMPER AND COIL SPRING ASSEMBLY

Faults to check are—

1. Lock-nut loose
2. Nut insufficiently tightened
3. Washer damaged
4. Nyloc nut loose
5, 8. Rubber bushes insufficiently compressed
6. Washer missing
7. Upper spring pan distorted
9. Washer damaged
10. Road spring weak or broken
11. Damper worn
12. Lower spring pan distorted
13. Collets incorrectly fitted
14. Bolt loose or worn
15, 17. Washers missing
16. Metalastic bush worn
18. Nyloc nut loose

CHART No. 24

SHOCK ABSORBER DEFECTS

Apart from defective shock absorbers, faults in springs (Chart No. 23), or incorrect tyre pressures, can cause poor suspension.

HYDRAULIC SHOCK ABSORBERS

Insufficient Damping Action — Check fluid level and top-up if necessary — Disconnect link from operating arm and pump arm up and down to expel air —
- Resistance to movement of arm erratic or non-existent — Examine and clean valves if easily accessible — If valves are clean, or if not accessible, shock absorber must be dismantled (see text) to check for internal faults —
  - Faulty rebound or compression valve.
  - Leaking piston(s).
  - Worn or scored cylinder(s).
  - Broken piston-operating mechanism.
- If resistance is uniform, adjust if possible (see text) — Check for mechanical faults —
  - Worn rubber bushes in arm and connecting link.
  - Operating arm loose on splines or cotter loose.
  - Shock absorber attachment bolts loose.

Excessive Damping Action — Check for too high fluid level: no air space left for expansion when warmed up — Check that correct grade of fluid is in use. Piston-type shock absorbers use thinner fluid than vane-type — If correct, check adjustment if accessible —
- Incorrect adjustment.
- Sticking valves.
- Fluid passages obstructed.
- Seized piston or rotor arm.

Leakage of Fluid — Ascertain point of leakage. If from gland, tighten if adjustable (see text) — If from flange or cover joints, tighten joints (see text) — If foregoing fails to cure leakage —
- Defective gland.
- Defective gaskets.
- Top cover or end caps distorted.
- Shock absorber overfilled.

Noisy Operation — Check for lack of fluid — Verify that rocker arm is not loose and that attachment bolts are tight. Also possible that rubber bushes in arm or link have worn through — If no external faults are evident —
- Internal faults: worn pistons or linkage.
- Broken parts in operating mechanism.

FRICTION-TYPE SHOCK ABSORBERS

Incorrect Damping Action — Check action of shock absorbers with one arm disconnected, preferably measuring resistance with spring balance —
- Insufficient friction with adjusting nut tightened up —
  - Glazed friction disks.
  - Friction disks worn out.
  - Spring weakened.
- If friction is adequate check action of hydraulic expander, if fitted. — If unsatisfactory, check for following faults (see text) —
  - Lack of fluid in reservoir.
  - Loose pipe line unions.
  - Air in pipe lines.
  - System not primed to correct pressure.
  - Grit on seating of ball valve in reservoir.
  - Defective expander.
  - Defective regulator.

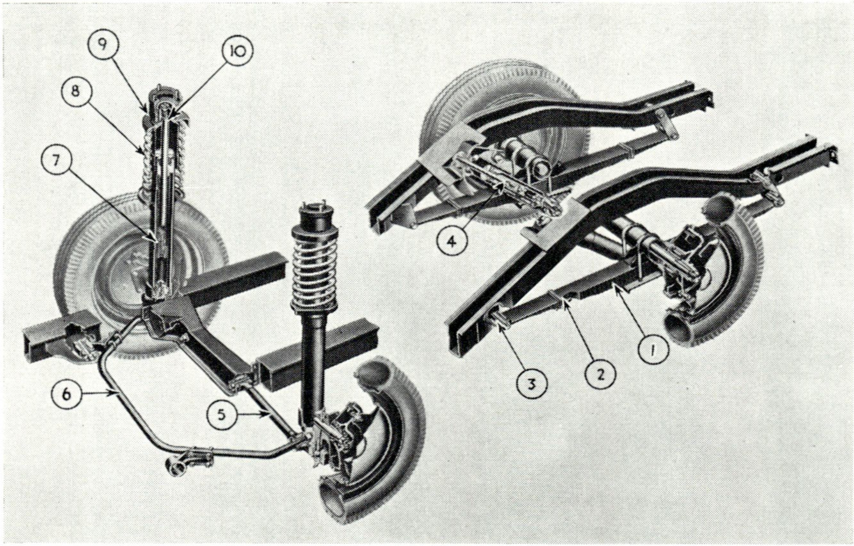

Fig. 105. Combination of Coil and Leaf Spring Suspension used on Recent Ford Models

1. Laminated spring
2. Rubber inserts
3. Rubber bushes
4. Shock absorber
5. Track control arm
6. Stabilizer bar
7. Suspension unit
8. Coil spring
9. Upper spring seat
10. Thrust bearing

shock absorber unserviceable. A modern shock absorber seldom has a useful life exceeding 20,000 miles; it may in fact have lost a great deal of its efficiency during a much shorter period of use, if the car is driven fast over bad roads. Many of the faults given in the last column of Chart No. 24, therefore, are included as being of interest, rather than as a basis for repair or rectification.

It is possible to check whether a telescopic damper has failed by removing it from the car and extending and compressing it through a full stroke by hand, taking care to keep it upright so that there is no risk of air entering the working chamber. If there is no resistance to the movement, or it is erratic, a new unit should be fitted. It is difficult to check the efficiency of a damper accurately in this manner, however—apart from detecting complete failure—since a large part of the damping control during normal running depends on the high-pressure valve, which comes into action only with rapid, repeated movements which cannot, of course, be obtained by hand testing.

A piston-type shock absorber can often be tested for failure as just described, without removing it from the car, by disconnecting the operating link and moving the arm through a full stroke by hand.

With either type of shock absorber, check the rubber bushes in the mountings or operating link for wear or contamination by oil. If the mountings are compressed by a threaded rod, check whether slackness can be eliminated by tightening the adjusting nut. If the mountings are worn or compressed, appreciable suspension movements may take place without operating the shock absorber, thus giving the impression that the dampers themselves are faulty.

CHAPTER XII

THE CHASSIS AND BODY

WHEN steering defects and rapid tyre wear cannot be traced to any of the faults covered in Chapters VIII-XI, it is possible that they may be caused by a distorted or twisted chassis frame, especially if the car has at some time suffered minor accidental damage. Major damage caused by a full-scale crash is, of course, unlikely to be overlooked, but the possibility of the chassis being slightly distorted by a blow on the dumb iron or front cross member, and similar parts, is not always given the attention which it deserves.

Although nowadays most conventional chassis frames are of very stiff construction, this type of chassis is more likely to suffer from general distortion than is the integral body-chassis form of construction, in which any damage is usually localized near the point of impact.

How to Check a Chassis. In either type of frame, however, it is a simple matter to make a quick check by placing the car on a clean, level surface, preferably a concrete floor, and suspending a plumb line from four fixed points on each side of the chassis. When semi-elliptic springs are fitted, the grease nipples in the eyebolts passing through the fixed ends of the springs, and through the chassis attachment points of the swinging shackles, will serve as reference points. On independent suspension systems suitable pivot pins in the chassis attachments of the suspension arms should be chosen.

At each point chalk an area on the floor beneath the plumb bob and carefully mark the position of the plumb bob with a pencilled cross. Having plotted all the points in this manner, push the car away and measure the diagonals between pairs of points. The measurements on two similar diagonals should agree to within about $\frac{1}{4}$ in. on the shorter diagonals and $\frac{5}{16}$ in. on the longer ones. The diagonals may be drawn in with chalk, as shown in Fig. 106. They should cross on the centre-line, and the arcs drawn from any two opposite points should also intersect on the centre-line.

Usually, when distortion of the chassis is revealed it will, of course, be necessary to remove the body in order to carry out repairs to the chassis. Localized damage to the dumb irons or the engine sub-frame can, however, often be put right with the aid of an hydraulic jack and wringing irons, if the amount of bending or buckling is not too great. Otherwise it will be necessary to heat the chassis members to allow them to be straightened without further damage, or it may be essential —and cheaper—to cut out the damaged section and weld in a new part, particularly with composite body-chassis construction. This work is outside the scope of this book, and is, in any case, a job for the specialist, in view of the welding skill required and the necessity for using special alinement and assembly jigs in many cases.

Checking Body Alinement. When body distortion is also suspected, the alinement of the basic structural members can be checked by the measurement of diagonals, a principle somewhat similar to that just described being used, and discrepancies in measurement being relied on to show up any distortion. A door opening may be taken as an example. A rectangle is constructed by marking a point at a given height on each door pillar. From these points a plumb line is dropped and in each case the door sill immediately below the plumb bob is marked.

To ensure that the upper marks are at the same height, it is best to hold a straight batten horizontally across the door opening at the required height, level it with the aid of a spirit level, and make chalk marks on the pillar against its upper or lower edge.

With a rectangle constructed on each door frame, measurements of pairs of diagonals can be compared. It is preferable to use an adjustable trammel, as shown in Fig. 107.

On all-steel bodies an appreciable amount of distortion can be corrected by pushing the offending member back into place with a body jack, but on coachbuilt bodies rectification is a task for the expert coachbuilder.

Tracing Body and Chassis Noises. Body and chassis noises, water leaks, draughts in the body, and similar troubles are often caused by faulty alinement of various parts, although the lack of alinement need not necessarily be caused by serious twisting or distortion of the body or chassis frame.

As will be seen from Fig. 108, for instance, the doors are

CHART No. 25 CHASSIS DEFECTS

Chassis Alinement Faults

- Make preliminary inspection for visible damage
  - Check alinement of chassis by measurement of diagonals, as described in text
    - Diagonals differ in length by amount which exceeds $\frac{1}{4}$–$\frac{5}{16}$ in.
      - Bent dumb irons.
      - Engine mountings damaged.
      - Side frame members bent.
      - Cross-bracing members buckled.
      - Rivets loosened or sheared.
      - Welds broken.
    - Diagonals within tolerances
      - Fault must lie elsewhere
        - Check front axle and steering gear (Chart No. 18).
        - Check suspension (Chart No. 23).
        - Check wheels, hubs and tyres (Chart Nos. 16 and 17).

CHART No. 26 CHASSIS LUBRICATION TROUBLES

Pressure-Gun Lubrication Systems

- Thoroughly clean nipple before applying grease gun Check for one of three faults
  - Lubricant does not enter nipple
    - Examine mating faces of nipple and grease gun. If undamaged:
      - Mating faces of gun and nipple are not of same pattern.
      - Nipple choked.
      - Ball valve in nipple stuck in closed position.
      - Oilway or pipe to bearing blocked.
  - Lubricant leaks from nipple
    - Unscrew nipple and soak in paraffin or penetrating oil
      - Ball valve held off seating by dirt or grit
      - Valve spring defective.
  - Lubricant enters nipple but does not reach bearing
    - On grouped lubrication systems check pipe lines for leakage
      - Chassis distribution pipe lines leaking.
      - Oil escapes from badly-worn bearing before reaching others served by same nipple.
      - Oilways blocked.
    - One bearing starved
      - Pipe line obstructed, leaking or broken
      - Restrictor choked.
      - Wrong size of restrictor pin fitted.

AUTOMATIC CHASSIS LUBRICATION SYSTEMS

- Check fluid level in supply tank. Then examine pipe lines. Determine amount of oil reaching bearings
  - Pipe line and restrictor faults
    - Group of bearings starved
      - Pipe line serving group obstructed, leaking or broken.
    - All bearings starved
      - System not primed with oil.
      - Main distribution pipe faulty.
      - Wrong size of restrictor in pump.
      - Pump defective (see below).
    - Excessive oil reaching bearings
      - Wrong sizes of restrictor pins in unions.
      - Pump delivery excessive (see below).
  - Pump faults
    - Foot pedal-operated pump
      - Pedal operated too frequently or not frequently enough.
      - If pedal remains depressed, filter in pump is clogged or main distribution pipe choked.
      - Faulty valves in pump—test with outlet unions sealed by tightly-fitting plugs: pedal should remain depressed.
    - Vacuum-operated pump
      - Check condition of regulator in suction pipe union. Replace at 10,000-mile intervals
        - Filter in regulator choked. Fit new regulator of same "rate" number.
        - Pump defective—check by disconnecting outlet pipe and accelerate and decelerate engine a number of times: few drops of oil should be expelled from outlet union.
        - If no pumping action, check filter.
    - Thermally-operated system
      - Expansion chamber incorrect distance from exhaust manifold—move closer to increase output; further away to decrease.
      - Valves faulty. Unscrew after disconnecting inlet and outlet pipes and wash in paraffin.
    - Engine lubrication pump system
      - Test regulator with outlet pipe disconnected.
      - If oil is not supplied with engine running at moderate speed, fit new regulator.

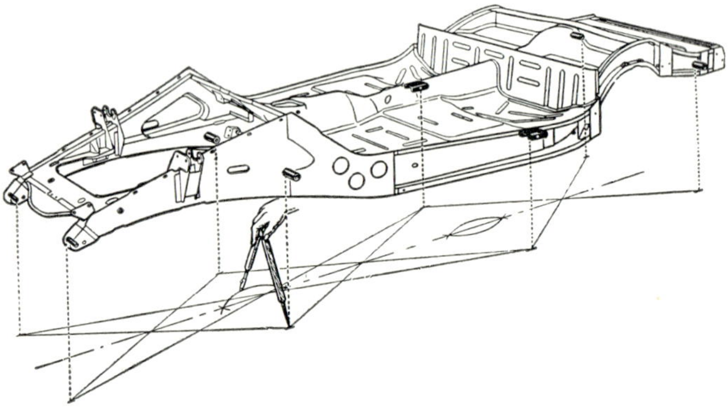

FIG. 106. A SIMPLE METHOD OF CHECKING A CHASSIS OR UNDER-FRAME BY PLOTTING FIXED POINTS ON THE FLOOR, AND VERIFYING THE DIAGONALS

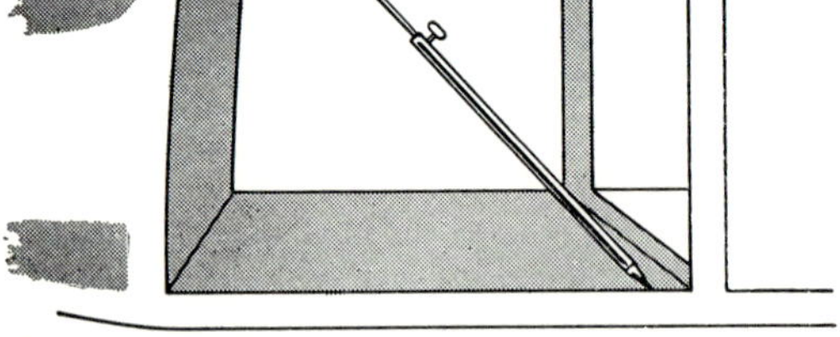

FIG. 107. CHECKING A DOOR FRAME DIAGONAL

frequent offenders, and metal-to-metal contact may take place at a number of points. The caption is self-explanatory, and Figs. 109 and 110 show methods of opening out or closing-in hinges very often used on all-steel bodies. The door itself can also be " set " to correct distortion on this type of body.

Curing Door Rattles. Nevertheless, drastic treatment is not always required. Attention to the door locks and anti-rattle

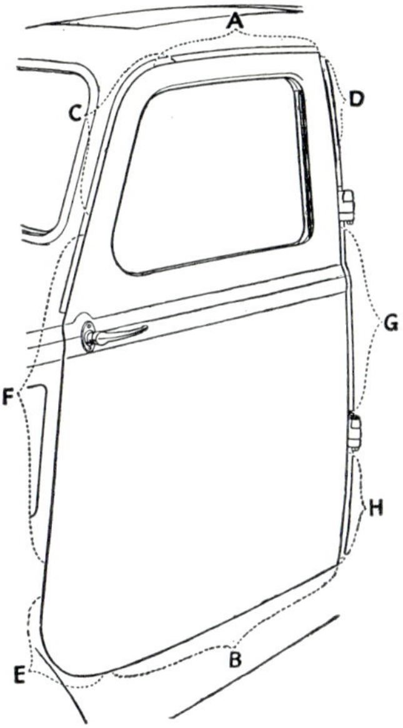

FIG. 108. POINTS AT WHICH METAL-TO-METAL CONTACT CAN TAKE PLACE BETWEEN DOOR AND FRAME, CAUSED BY THE FOLLOWING FAULTS

- *A* Hinges are too low on door
- *B* Hinges are too high
- *C* Door dropped—pack out lower hinge or bend out
- *D* Top hinge closed in too far—pack out or open out by bending
- *E* Door has dropped—pack out lower hinge
- *F* Incorrectly fitted hinges or dovetail
- *G* Opposite of *F*—re-set both hinges
- *H* Close in top hinge and open out bottom hinge

devices is often all that is needed. In most doors the lock striker plate is adjustable, having serrations on its rear face which match serrations on a plate attached to the door frame. After the attaching screws have been slackened, the striker plate can thus be moved inwards by one or two serrations, causing the door to close more firmly.

The wedge-shaped locating devices on the door and pillar, which may be in the form of rubber wedges, tapered dowel pins, metal wedges, or spring-loaded dovetails, are generally provided with adjustments which allow the two members to be moved closer to one another, restoring the correct positioning action. When rubber wedges become worn, replacement is the only cure, but a further lease of life can often be obtained from them by changing them over or turning them round in order to bring an unworn section into use.

When the door is correctly fitted and located, it is probable that any rattles which appear to emanate from it originate in the window glass or winding mechanism. It is a simple

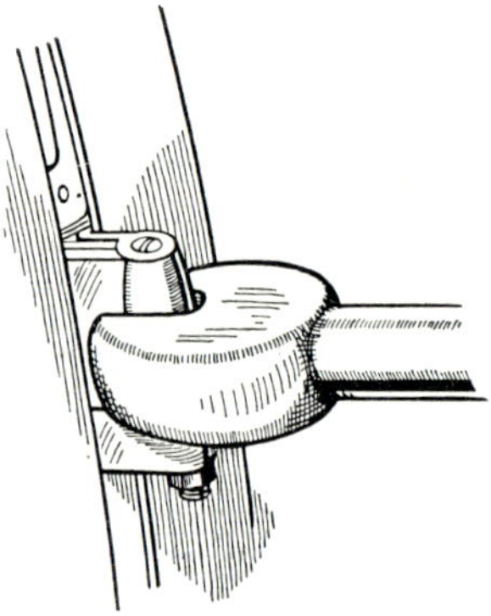

Fig. 109. Closing-in a Hinge

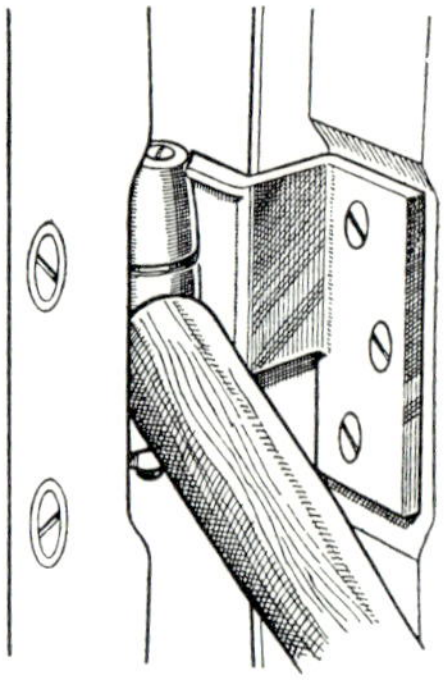

Fig. 110. Opening-out a Hinge

matter to eliminate the glass as a possible offender by temporarily wedging it ; if the rattle is still present, the winding mechanism is probably at fault.

Window Winding Mechanism. To obtain access to the winding mechanism, or to remove the window glass in order to fit new felt or rubber strips to its channels, the interior trimming of the door must be removed. The method of attaching the trimming panel varies. Preliminary work required will usually include removal of the window capping or frame. This may be secured by small screws, or by spring clips, in which event it must be gently levered away, taking care not to fracture a plastic capping. The interior door handle and window winding-handle will also have to be removed. Among methods of attaching the handles to their spindles are : a pin passing through the spindle, which can be pushed out when the escutcheon plate has been pressed back;

a spring wire clip incorporated in the handle, which must be pressed back by passing a bradawl through the base of the handle ; a magnetized key ; or a spring-loaded fixing in the escutcheon plate which is disengaged by rotating a tab on the underside of the handle.

The trimming panel may be secured by small screws or tacks, hidden by a trimming tape. On all-steel bodies spring clips are used instead, and the panel may be carefully levered

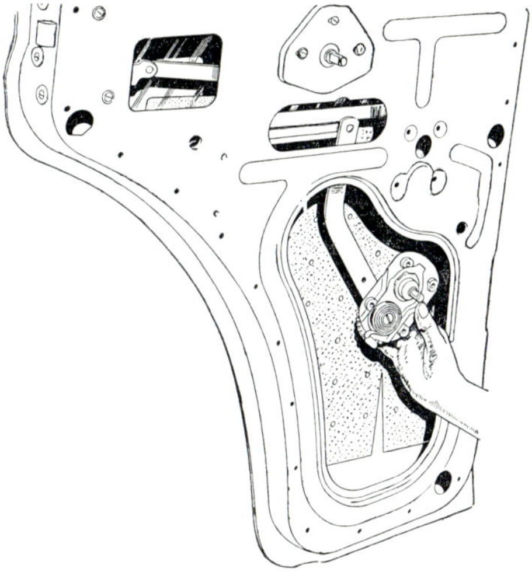

FIG. 111. REMOVING THE WINDOW-WINDING MECHANISM ON A HILLMAN MINX

away by inserting a screwdriver at intervals around its edge.

The winding mechanism will now be revealed, and faults such as a broken balancing spring, worn rollers and guides and worn or stripped gear teeth will be evident. The method of removing the winding mechanism, window glass, and channels varies considerably in different cars, but will generally be evident on inspection, followed by a tentative trial of what appears to be the most obvious method.

It is usually necessary to remove the wooden block or rubber stop which limits the downward movement of the glass, so that when this is lowered beyond the normal point the rollers or buttons on the lifting arms can be brought clear

CHART No. 27

BODYWORK TROUBLES

- BODY NOISES — Diagnosis is usually rendered difficult by fact that noises are often transmitted by body and chassis members to points remote from the actual sources. Localize noise as far as possible:
  - At front of car
    - Rattles
      - Loose bumper or number plate fittings.
      - Head lamp or side lamp rims, glasses or reflectors loose.
      - Loose tools.
      - Loose wheel nave plates.
      - Mudguard bolts slack.
      - Loose battery mounting.
      - Radiator grille loose.
      - Radiator mounting worn.
      - Radiator tie rods incorrectly adjusted.
      - Suspension faults (Chart Nos. 23 and 24).
    - Squeaks
      - Any of above faults.
      - Worn bonnet tapes.
      - Engine noise: from distributor shaft, dynamo bearings, fan belt (Chart No. 4).
  - Noises from doors
    - Worn locks.
    - Worn or incorrectly adjusted buffers and dovetails.
    - Worn or incorrectly adjusted hinges.
    - Worn or broken parts in window winding mechanism.
    - Distorted door.
  - Rattles or squeaks from windows
    - Felt or rubber channel strips worn or hardened.
    - Anti-rattle device defective.

- Amidships and inside body
  - Other causes of rattles and squeaks
    - Worn steering column bearing.
    - Rattle in steering hub control plate.
    - Vibration in gear lever.
    - Worn seat slides.
    - Loose instrument panel or facia board.
    - Worn speedometer drive cable.
    - Loose floor boards or inspection panels.
    - Running boards loose.
    - Body bolts loose.
    - Defective silencer and exhaust pipe supports.
    - Worn brake rod pivots.
- From rear of body
  - Spare wheel insecurely mounted or cover loose.
  - Luggage compartment lid not closing properly.
  - Bumpers loose.
  - Loose tools or other items in luggage compartmen
  - Defective exhaust pipe brackets.
  - Suspension faults (Charts Nos. 23 and 24).

Water Leakage and Draughts

- At doors and windows
  - Worn sealing strips or channels.
  - Badly-fitting doors.
- At windscreen and scuttle ventilator
  - Worn sealing rubber.
  - Sealing strip not cemented in place.
  - Windscreen incorrectly adjusted.
  - Defective seals at wiper arm spindles.
  - Tension of ventilator control incorrectly adjusted.
  - Scuttle ventilator water drain pipe blocked (if fitted).
  - Windscreen frame distorted.
  - Ventilator or flange distorted.
- At sunshine roof
  - Roof channels improperly sealed.
  - Drain pipes obstructed.
  - Roof warped or not closing properly.

of the channels, allowing the glass to be lifted upwards and outwards. On the rear door windows, however, it is often necessary to remove the channels in which the glass slides.

Other Body and Chassis Noises. Little need be said in amplification of the various sources of squeaks and rattles covered in Chart No. 27. Successfully tracking down an elusive noise often calls for a considerable degree of patience, one possible cause after another being systematically eliminated.

The most likely faults are covered in the chart, but it is, of course, impossible to forecast every defect which may give rise to an irritating noise. The services of an assistant, who will be able to carry out tests while the car is being driven, will be found invaluable. It is also worth while stationing an assistant by the roadside while the car is driven past, in order that external rattles caused by bumpers, number plates, lamps and similar fittings, may be detected.

Curing Water Leaks and Draughts. Causes of entry of water into the body and the occurrence of draughts have been grouped for convenience in Chart No. 27, since both can be caused by an ill-fitting joint. It must not be assumed, however, that both faults will necessarily appear together. Water leaks, in fact, are often among the most persistent troubles to eradicate, obstinately reappearing when apparently every likely crack or opening has been sealed.

In amplification of the chart one or two points are worth bearing in mind. The majority of opening windscreens are provided with adjustable hinges, for instance, which allow the screen frame to be drawn more closely into contact with the sealing rubber. Slotted hinge plates may be concealed beneath the interior capping at the top of the screen, or the hinges may be provided with countersunk adjusting screws, locked by hexagonal nuts. Adjustment of the screen should not be overdone, however, because of the risk of distorting the screen frame and cracking the glass.

On fixed windscreens, leakage past the rubber seal can invariably be prevented by carefully cementing the seal in place with a suitable cement, such as Bostik, the lip of the rubber being raised by a knife to allow the cement to be injected well into the joint. Glazing compound is useful to seal quarter window glasses in their frames, to stop up small cracks, and similar applications.

Sunshine roofs should not leak provided that the channels are adequately drained ; the small drainage holes, and the tubes fitted on modern cars to carry the water away below chassis level often become choked with dust and other foreign material, so that an occasional check-over is advisable.

It is usually possible to clear obstructed drainage tubes by using a length of flexible wire, taking care, of course, not to jam the obstructing matter solidly in the tube by using excessive force.

If it is necessary to remove the sunshine roof for attention to the channels or for any other reason, the advice of the manufacturer should be sought, as the method of freeing the roof from its guides varies on different cars, and the use of force or injudicious leverage is liable to damage the roof or its channels.

CHAPTER XIII

THE INSTRUMENTS AND AUXILIARIES

THE methods of tracing trouble in faulty instruments and auxiliaries outlined in Charts Nos. 28 and 29 should be found to be quite straightforward. The majority of the tests relating to instruments must aim at determining whether the instrument itself is at fault, or whether the trouble is due to a faulty electrical circuit, a choked or damaged pipe line, or a faulty driving mechanism, depending on the instrument in question. Actual repair of the instrument is seldom possible : special tools are generally required, and a test rig will be needed to check the calibration aftre repair. The only remedy, therefore, is to fit a new instrument or to return the defective one to the makers for repair.

If a replacement instrument can be borrowed for testing purposes, fault-tracing is, of course, considerably simplified, since if the replacement functions satisfactorily, it is obvious that the original must be faulty. Conversely, if incorrect readings are obtained with the replacement connected up, the fault must lie elsewhere than in the instrument. As it is not always possible to take this useful short cut, however, the chart has been arranged so that fault-tracing can be carried out with the minimum of equipment.

Obtaining Access to the Instruments. When it is necessary to obtain access to the rear of the instruments for fault-tracing purposes or in order to remove them, it may be advisable to detach the complete panel from its mountings so that it can be moved forward and tilted to render the instruments more accessible. This is generally advisable on panels on which a number of instruments are grouped on one dial, since the close proximity of the various terminals is apt to lead to mistaken connexions and accidental short-circuits when testing.

On some cars a detachable panel on the engine bulkhead can be removed, rendering instantly accessible the back of the instrument panel. In most cars, however, it will be necessary to remove part of the facia board before the screws securing the panel can be reached. Before moving the panel, disconnect one battery lead to prevent short-circuits, and make sure that sufficient slack exists in the various leads and pipe lines.

While the speedometer driving-cable clamp and oil-gauge pipe union can be unscrewed, the small-bore tubing, connected to a water temperature gauge (if fitted), must on no account be damaged or any attempt be made to detach it from the gauge. Sufficient slack is generally available in the form of a coil of tube behind the dashboard, but if it is not, the temperature bulb must be removed at the engine or radiator union, and the tubing released from its clips and carefully fed back into the driving compartment.

The method of securing the instruments to the panel varies. On modern cars the instruments are generally attached by screws or by retaining plates which are secured by screws. On older cars a strip-steel stirrup, secured to the instrument by two nuts, may be employed, or a coil spring surrounding the instrument, and retained by a flanged locking ring, may be used.

If any electrical connexions are disconnected, a note should be made of the terminal from which each is removed ; the use of colour-coded insulation is almost universal today, so that reconnexion is a simple matter if a wiring diagram is available or a key is jotted down at the time of removal. Otherwise the leads should be suitably labelled.

Steering Column Switch Assemblies. When attention is required by the switches or horn push mounted on the steering wheel boss, the switch assembly may be retained by two or more screws passing through the underside of the boss. Alternatively, the assembly may be carried on a tube which passes through the centre of the steering column, and which is secured at the base by a split clamp outside the steering gearbox. Sometimes the tube is screwed into the base of the steering gearbox itself. In either case it is generally necessary to drain the oil from the steering gearbox before withdrawing the tube and switch assembly.

Another arrangement which may be encountered is a switch box attached to the base of the steering gearbox, and operated by a rod and tube passing up through the column. Dismantling this latter type of switch is therefore a simple matter.

Before withdrawing the more usual type of switch assembly

CHART No. 28 DIAGNOSING DEFECTIVE INSTRUMENTS

PETROL GAUGE READINGS INACCURATE

- A.C. AND SMITHS GAUGES HAVING TWO ELECTRICALLY-OPERATED MAGNETIC COILS IN INSTRUMENT (*see* page 237)
  - Gauge needle remains at "Empty"
    - Disconnexion or break in wire from ignition switch to gauge.
    - Wire between gauge and tank unit earthed.
    - Gauge casing not earthed.
    - Connexions to gauge reversed.
    - Float punctured or saturated.
    - Float arm sticking.
    - Faulty gaure.
  - Gauge needle remains at "Full"
    - Disconnexion or break in wire between gauge and tank unit.
    - Tank unit not earthed.
    - Open circuit in resistance or wiper arm in tank unit
    - Float arm sticking.
    - Faulty guage.
  - To determine whether instrument or tank unit is at fault, disconnect wire from "T" or "Sender" terminal on gauge. Switch on ignition. Gauge should read "Full." With wire still disconnected, earth "T" or "Sender" terminal. Gauge should read "Empty." If these tests are satisfactory the instrument is serviceable.
- INSTRUMENTS OPERATED BY BI-METAL STRIP (*see* page 238)
  - No reading on gauge
    - Voltage regulator not supplying 10 volts.
    - Disconnexion or break in wiring between regulator and gauge.
    - Faulty gauge or transmitter, check by substituting serviceable unit.
  - Intermittent readings
    - Faulty voltage regulator.
    - Faulty indicator or transmitter.
    - Loose connexions in wiring.
  - High or low reading on gauge
    - Faulty voltage regulator.
    - Faulty gauge.
    - Faulty transmitter.
    - Wiring short-circuiting to earth

- **Incorrect Oil Pressure Gauge or Warning Light Indications** — If possible test by connecting accurate gauge to engine union. If correct pressure is shown, original trouble lies in gauge or pressure switch
  - **Conventional Bourdon Tube Gauge** — Check for leakage of oil from gauge
    - Oil leakage — Bourdon tube punctured or soldering faulty.
    - No oil leakage
      - Bourdon tube strained.
      - Oil gauge pipe line choked.
  - **Electrically-operated Gauge** — Operation is similar to fuel or temperature gauges actuated by bi-metal strips; Fault tracing as in chart above, but substitute "Engine Unit" for "Tank Unit" or "Temperature Transmitter," "Low Pressure" for "Empty," "High Pressure" for "Full."
  - **Oil Pressure Warning Light Switch**
    - Light continues to glow when oil pressure is adequate — Disconnect wire from engine pressure switch terminal
      - Light goes out — Faulty pressure switch.
      - Light continues to glow — Short circuit to earth in wire between pressure switch and warning light.
    - Light fails to glow when ignition is switched on, but before engine is started — Check bulb in warning light. If sound — Short-circuit pressure switch terminal to earth without removing wire
      - Light glows — Faulty pressure switch.
      - Light fails to glow — Disconnexion in wire from ignition switch to warning light or light to pressure switch.

- **Speedometer and Revolution-indicator Faults**
  - **Needle Remains at Zero** — Check driving cable by disconnecting at instrument and observing whether shaft rotates when car is in motion (speedometer) or when engine is running (revolution indicator)
    - Shaft rotates
      - Shaft incorrectly engaged with instrument spindle.
      - Driving key missing, or thrust washer omitted.
      - Mating surfaces of shaft and spindle worn.
      - Defective instrument.
    - Shaft does not rotate
      - Driving cable broken.
      - Driving key or socket at gearbox end of shaft sheared or worn.
      - Driving gears stripped.
      - Undue friction between cable and casing—lubricate.
  - **Wavering and Erratic Readings** — Remove driving cable from casing and inspect for lack of lubrication, stretched cable, and similar faults
    - Sharp bends in casing.
    - Worn or stretched cable.
    - Belt drive slipping (belt drive is unusua nowadays).
    - Faulty instrument.
  - **Steady but Inaccurate Readings** — Many speedometers are calibrated to read about 10 per cent fast at speeds in excess of 35–40 m.p.h. — If error is excessive
    - Faulty speedometer
    - Wrong sizes of tyres on rear wheels.
    - Rear axle ratio altered from standard, necessitating recalibration of speedometer.

CHART No. 28 (continued)

- **Noisy Operation** — Tapping and whirring noises are usually caused by an unlubricated or faulty driving cable but may originate in instrument — Lubricate inner cable. If noise continues: — Drive car or run engine with cable disconnected from instrument
  - Noise continues — Faulty cable.
  - Noise ceases — Faulty instrument.

Moving-coil Revolution-indicator Faults — Test generator on engine with AC voltmeter. Reading should usually be approx. 1 volt per 100 r.p.m. of generator driving shaft
- No reading — Faulty generator.
- Satisfactory reading
  - Faulty indicator.
  - Defective wiring between generator and indicator.

Electronically-operated Revolution Counters — Check that connexions to ignition circuit are in accordance with manufacturer's diagram — If correct, check terminals for cleanliness and tightness — If connexions are sound — Faulty instrument carried on the steering wheel boss it will, of course, be necessary to disconnect the leads passing through the column at the nearest junction box. One is usually located close to the steering gearbox. A practical tip when a lead must be renewed is to attach the new lead securely to the end of the existing one, so that it is drawn into place in the tube as the old lead is withdrawn.

Direction Indicators. The faults likely to be experienced with semaphore-type indicators are covered in Chart No. 29. A similarly detailed chart is not required in the case of modern flashing indicators, since the system is much simpler; fault-tracing is confined to checking the indicator and warning lamp bulbs, bulb holders, wiring and switch.

The flasher unit itself, within a small, sealed container, consists of a switch which is operated automatically by the alternate heating and cooling of an actuating wire, and a small relay which flashes the warning light on the instrument panel or in the operating switch. As the flasher unit is not repairable, the only practicable test is by substitution of one that is known to be serviceable.

On some earlier cars, the same pair of bulb filaments is used both for the stop light and the rear direction indicators. As it is essential that response to the flasher unit shall take precedence over simultaneous application of the brake light switch, additional contacts may be included in the direction-indicator switch or a special double relay may be fitted. This relay should be checked in the case of failure of either the stop light or the indicator lights.

Stop Light Switches. A further practical point is that although the majority of stop light switches are mechanically operated by being coupled to the brake pedal, on hydraulic braking systems the switch may be operated by hydraulic pressure. If the switch is removed for attention, some fluid will be lost from the system, so that after refitting the switch the supply tank should be topped-up and the brake system "bled" to expel air, as described in Chapter X.

Windscreen Wipers. The faults detailed in Chart No. 29 should need little amplification. A point worth checking when vacuum-operated wipers are fitted, however, is that the tendency of the wipers to slow down or stop when the throttle is held open for any appreciable period—an inherent drawback

CHART No. 29

DEFECTS IN AUXILIARIES

- **Semaphore-type Direction Indicator Faults**
  - **Arm Fails to Rise** — Check ammeter reading. If not fitted, carry out both series of tests
    - Ammeter shows discharge when direction indicator switch is operated. — Arm jammed in case or recess. Do not attempt to prise arm out with switch off. Rock arm in case and apply pressure at top of arm.
      - Arm bent.
      - Arm returns too far into case or recess—slightly lever up metal tongue carrying rubber buffer.
      - Mechanism binding due to lack of lubrication—lubricate pivots and linkage sparingly with thin oil.
    - No discharge on ammeter — Electrical disconnexion. Check fuse. If not blown
      - Faulty switch.
      - Frame or casing of indicator improperly earthed.
      - Defective solenoid.
  - **Bulb does not Light when Arm Rises** — Inspect bulb. If in order:
    - Bulb improperly positioned in carrier.
    - Bulb carrier does not make good contact with arm or clip.
    - Wire to bulb broken.
    - Switch contacts at inner end of arm dirty.

Flashing indicator faults—*See* page 233

- **Electric Wiper Motor with Wound Armature** — Check current taken. If no ammeter fitted test with separate ammeter or make visual examination for faults in last column
  - No current passing
    - Disconnexion or broken wire in feed or earth return circuit. Fuse blown.
    - Switch contacts dirty—clean with fine emery or glass paper.
    - Loose or broken connexions in motor.
  - Field current only taken: $\frac{1}{2}$–$1\frac{1}{3}$ amps.
    - Brushes worn or sticking.
    - Oil, grease or carbon on commutator.
    - Open circuit in armature.
  - Normal or excessive current taken—2–$4\frac{1}{2}$ amps. or more. Motor overheats
    - Commutator dirty.
    - Armature bearings binding.
    - Brush gear short-circuited.
    - Excessive friction on wiper spindle.
    - Excessive blade pressure.
    - Defective gears or linkage.

WINDSCREEN WIPER DEFECTIVE

- **INDUCTION TYPE ("SPIN-START") MOTOR** — Check current taken or make visual examination
  - No current taken and no impulses felt when starting knob is turned
    - Switch contacts dirty or incorrectly adjusted.
    - Disconnexion or broken wire in feed or earth circuit. Fuse blown.
    - Contact breaker contacts not closing or faces dirty.
    - Open circuit in field winding.
    - Loose or broken connexions in motor.
  - Motor runs sluggishly
    - Contact breaker gap incorrect—adjust by trial.
    - Excessive friction in spindle or gearing.
    - Excessive blade pressure.
  - Rotor magnetically locked between pole pieces
    - Contact gap incorrect.
    - Fibre cam incorrectly set—cam is tight push fit on spindle and may be rotated to alter timing of points.
- **VACUUM-OPERATED WIPERS** — Check suction pipe between induction manifold, wiper and vacuum reserve tank (when fitted) — Check action of valve in vacuum reserve tank and of wiper control valve — If vacuum system is in order, fault must lie in wiper
  - Air leaks past pistons or paddle—lubricate or renew.
  - Throw-over valve in wiper sticking or valve spring stretched or bent.
  - Spindle binding or bent.
  - Excessive blade pressure.
- **MECHANICALLY-OPERATED WIPER FAILS TO OPERATE** — Disconnect driving cable outer casing and observe whether driving cable is rotating with engine running and wiper control engaged
  - Driving cable rotating
    - Fault in worm wheel or worm gear driving spindle.
    - Spindle sheared or loose in worm.
  - Driving cable not rotating: Disconnect engine end of cable and observe whether engine-driven gear shaft is rotating
    - Engine-driven gear shaft rotating
      - Broken driving cable.
      - Dog clutch worn or spring weak.
      - Operating control incorrectly adjusted.
    - Engine-driven shaft not rotating
      - Faulty camshaft gear or driven gear.

CHART No. 29 (continued)

- **Electrically-Operated Horns**
  - **Horn does not Sound** — Check external wiring and fuse — If in order:
    - Defective switch.
    - Broken wire in steering column.
    - Relay (where fitted) defective or incorrectly adjusted.
    - Burnt contact points in horn or relay. (See below.)
    - Defective or incorrectly adjusted horn.
  - **Horn Sounds Continually** — Make sure switch is not sticking:
    - Defective switch.
    - Short-circuit in switch wire.
    - Defective relay or incorrectly adjusted relay.
  - **Incorrect Note** — Check mounting of horn. If of flexible type, make sure horn is free to vibrate. If rigid mounting, check tightness of securing bolts — Check dynamo charging voltage:
    - Too high or too low dynamo voltage.
    - Excessive voltage drop in circuit.
    - Loose parts in horn.
    - Horn incorrectly adjusted: set contact gap and air gap to maker's recommendation.
    - Matched horns not tuned to chord.
    - Defective horn.

- **Air-Operated Horns** — Check that current is reaching horn air compressor motor
  - If no current:
    - Faulty horn push.
    - Faulty relay.
    - Disconnexion or short-circuit in wiring.
  - Current reaching compressor motor
    - Motor does not run — Faulty motor.
    - Motor runs:
      - Faulty compressor.
      - Blockage or disconnected pipe line to horn(s).
      - Faulty horn(s).

with this type of wiper which is not entirely overcome by the fitting of a vacuum-reserve tank—may be aggravated by a faulty vacuum-boost pump, which, on a number of recent models, forms an integral part of the mechanical petrol pump.

Two points are often overlooked with modern electrically-operated wipers. Some circuits incorporate a current-temperature operated switch which protects the wiper motor from serious overloading if the blades are prevented from moving by, for example, packed snow or an accumulation of ice on the windscreen. The current supplied to the motor will then be interrupted and will not be restored until the motor has cooled down, which may take an appreciable time—say, about ten minutes. The thermostatic switch is either located inside the wiper motor or is connected into the supply cable.

The second point concerns the self-parking arrangement adopted on many modern wipers. On some earlier Lucas DR2 wipers, the position in which the arms automatically come to rest when the wiper is switched off can be adjusted by slackening the screws retaining the cover of the wiper gearbox and rotating the cover slightly in the appropriate direction. In an alternative design, an adjusting nut projects from the gearbox, close to the point at which the cable rack enters it.

Thermostatic Interrupters. Mention of the thermostatic switch in the wiper circuit prompts the reminder that on some modern cars—notably the Vauxhall range—a thermostatic interrupter is fitted instead of conventional fuses. When a circuit is overloaded, the excess current passing through the interrupter switch causes contacts to open and break the circuit. The interrupter cools and contact is restored, only to be broken almost immediately. Thus the vibrating contacts prevent the passage of excessive current in the case of a short-circuit at some point in the system, while still permitting sufficient current to flow to, say, the head, side and tail lamps, to allow the car to be driven to the nearest service station.

Fuel Gauges. Two types of fuel gauge are in common use. In the type shown in Fig. 113, a small armature attached to the pointer is attracted by two electro-magnetic coils in the gauge. The current flowing through the coil which is connected to the "T" or "Sender" terminal on the gauge fluctuates in accordance with the position of a contact in the tank unit, which slides over a resistance as the float of the unit rises and falls. The pointer thus moves towards "Empty" or "Full." This type of gauge can be tested as indicated in Chart No. 28.

An alternative type of gauge (Fig. 114) which is widely

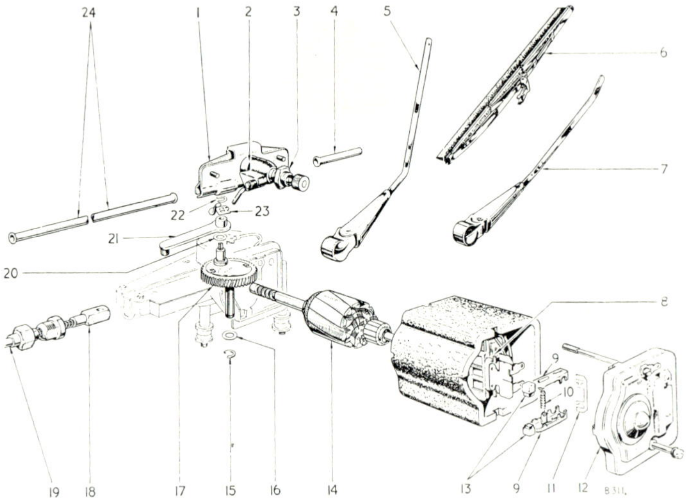

Fig. 112. Possible Windscreen-wiper Faults

1. Wheel-box worn or damaged
2. Jet and bush assembly worn
3. Nut loose
4. Rigid tubing—right-hand side—damaged or incorrectly assembled
5–7. Wiper arms and blades require renewal
8. Field coil assembly defective
9. Brushgear damaged
10. Tension spring weak
11. Brushgear retainer damaged
12. End-cover damaged
13. Brushes worn
14. Armature windings defective, commutator dirty or scored
15. Circlip broken
16. Washer missing
17. Final-drive wheel worn or damaged
18. Cable rack worn or damaged
19. Rigid tubing—left-hand side—damaged or incorrectly assembled
20. Spacer missing
21. Connecting rod worn
22. Circlip broken
23. Parking switch contact dirty or incorrectly adjusted
24. Rigid tubing—centre section—damaged or incorrectly assembled

used on recent cars incorporates a bi-metal strip in the instrument which is surrounded by a heating coil. Current, supplied at 10 volts by a small voltage regulator, flows through the coil to the tank unit and to earth through a variable resistance operated by the float. The current flowing through the coil heats the bi-metal strip, causing it to deflect to a varying extent and thus move the needle across the scale. Fault tracing with this type of instrument is also covered in Chart No. 28.

Engine Temperature Gauges. In the mechanical type of temperature gauge the needle is operated by a Bourdon tube which is connected by a capillary tube to a temperature-sensitive bulb in the engine cooling system in the case of a water-temperature gauge or in the oil sump in the case of an oil-temperature instrument. The whole system is filled with

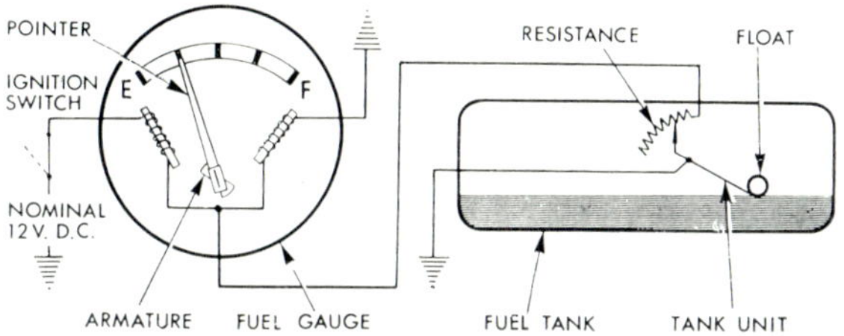

FIG. 113. CIRCUIT DIAGRAM FOR FUEL GAUGE OPERATED BY MAGNETIC COILS IN INSTRUMENT

ether. As the temperature rises in the bulb the ether expands, creating pressure which is registered by the gauge as temperature.

There is little that can be done to correct faults with this type of gauge, since failure is almost invariably due to the capillary tube being kinked or cracked, or to leakage occurring at one of the soldered joints. Refilling the gauge with ether and recalibrating it calls for expert knowledge and special equipment.

The more common type of temperature gauge found on modern vehicles is electrically-operated by current flowing through a coil surrounding a bi-metal strip, as in the case of the petrol gauge described above. Fault-tracing with this type of gauge is therefore on similar lines to that described for the fuel gauge in Chart No. 28.

An earlier type of Smiths electrically-operated temperature gauge receives current directly from the battery instead of from a voltage regulator. This pattern can be recognized by the fact that when the ignition is switched off the pointer drops instantly to "Cold." The pointer is controlled by a pair of electromagnetic coils as in the case of one of the designs of petrol gauge described earlier. The current flowing through the coil which is connected to the "T" or "Sender" terminal on the gauge is regulated by a transmitter bulb which contains a pellet of material, the electrical resistance of which varies with its temperature.

Fault-tracing with this type of gauge is confined to checking that battery current is reaching the "B" terminal on the instrument, that the wire connecting the "T" terminal to the

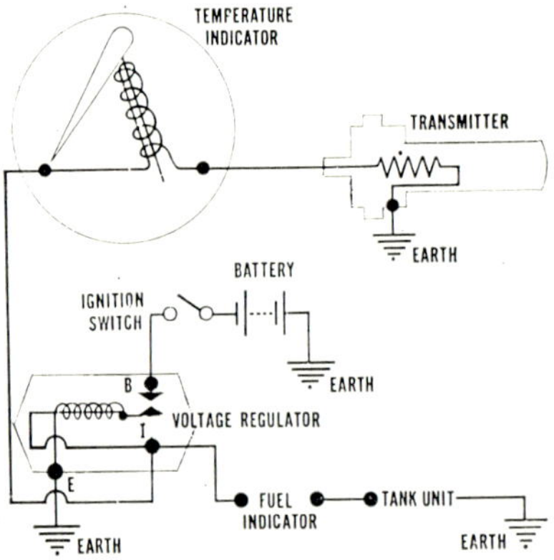

FIG. 114. CIRCUIT FOR THERMALLY-OPERATED TEMPERATURE GAUGE

The instrument is supplied with current at 10 volts by the voltage regulator, which also serves the fuel gauge (Fig. 115)

transmitter bulb is sound and that the instrument casing is properly earthed.

If these points are in order, to determine whether the instrument or the transmitter is at fault, disconnect the wire from the "T" terminal and switch on the ignition. The indicator should read "Cold." With the wire still disconnected from the "T" terminal, connect one lead from a 12-volt, 6-watt test bulb to the terminal and earth the other lead from the bulb holder. With the ignition switched on the indicator should read "Hot." In some cases the test can be made by earthing the gauge terminal directly, but the use of a test lamp prevents any risk of burning out the gauge windings.

Electric Clocks. When a Smith's electric clock is fitted it may be fitted with a rectifier. The normal, unrectified clock has a green moulding on the live connexion at the rear of the case. A rectified clock has a black moulding on this terminal. It is essential to maintain the correct polarity of a rectified clock. The case is therefore stamped —VE for positive-earth systems: i.e., the negative live lead is connected to the clock terminal. For negative-earth systems the case is stamped

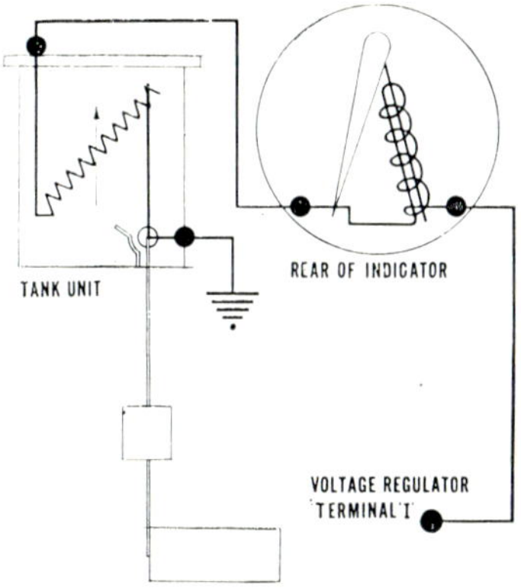

FIG. 115. CIRCUIT FOR THERMALLY-OPERATED FUEL GAUGE
The circuit for the voltage regulator is shown in Fig. 114

+VE and the positive lead is, of course, connected to the clock terminal.

Regulating Clocks. Although repair of a spring-wound or electric clock is a matter for a specialist, regulation of the clock if it should be gaining or losing time is a simple matter. On sliding aside a dust-cover a regulating lever similar to that used on domestic clocks may be revealed, or a gear-wheel type of regulator may be fitted. Movement of the latter type by one tooth towards " S " or " F " will generally result in the clock losing or gaining one minute per day. Another type of regulator takes the form of an adjusting screw on the rear or on the front of the clock case.

Checking Speedometer Accuracy. An obvious method of checking the accuracy of the speedometer is to drive at a steady speed, timing the car accurately as it passes two points that are known to be an exact distance apart. In practice, however, there are many possible sources of error when a test of this type is carried out on the road. For accuracy the test should be made over a distance of at least a quarter mile or half a mile, and the likelihood of being able to maintain an exact speedometer reading of, say, 40 m.p.h. over the whole distance is remote. When possible errors in handling the stop watch, and in the distance apart of marks such as milestones, are taken into consideration, it will be evident that the figures obtained can only be approximate.

A fairly accurate check can be made, however, by running the car with the rear wheels supported clear of the ground, the condition of the speedometer driving gears and cable having first been checked as detailed in Chart No. 28. In this check use is made of the mileage recording mechanism of the speedometer. This, being positively driven by gearing, is not subject to fluctuating or inaccurate readings, as is the speed indicating needle, which may be actuated by a centrifugally operated weight, by magnetic flux, or by a chronometric mechanism.

The mileage recorder should first be checked for accuracy by driving the car over a considerable distance, making use of points that are known to be a given number of miles apart, the distances preferably being based on ordnance survey figures rather than on old milestones, which are often inaccurately placed. The test should be made at a speed as steady as possible, to reduce the effect of slight wheelspin, and any error in the mileage reading, due to this and to other factors, such as any variation in tyre diameter, should be noted.

Next, with the rear wheels clear of the ground, and top gear engaged, adjust the engine speed until the speedometer needle is steady on a given figure. As an example 30 m.p.h. may be selected, in view of the importance, from the legal point of view, of accurate readings at this speed. A stop watch should be started when a unit figure registers on the mileage-recording indicator, and should be stopped at the instant that the next unit appears. At an indicated speed of 30 m.p.h. the time to register one mile on the mileage recorder, should, of course, be exactly two minutes. If any error exists in the mileage recorder this must be taken into account, although in most cases the error is negligible when related to a distance of one mile.

If the time taken is greater than two minutes, the speed-indicating mechanism is *fast*; if less than two minutes is required, the speedometer reads *slow*. As the latter is the most likely fault on an instrument which has seen considerable service, and may lead to legal action in a speed restriction area, the following figures, quoted to the nearest mile an hour, may be useful—

Time Taken to Register One Mile at Indicated 30 m.p.h.	Actual Speed
2 min. 0 sec.	30 m.p.h.
1 ,, 56 ,,	31 ,,
1 ,, 52 ,,	32 ,,
1 ,, 49 ,,	33 ,,
1 ,, 46 ,,	34 ,,
1 ,, 43 ,,	35 ,,

It must be remembered that an error at any given speed will not necessarily be constant at other speeds. Many speedometers are accurate at 30 m.p.h., slightly fast at 40 m.p.h., increasingly fast at 50-60 m.p.h., and sometimes as much as 10 m.p.h. fast at speeds in the region of 65-70 m.p.h. As this is a question of calibration by the maker, in conjunction with the car manufacturer, the fault cannot be corrected on a particular speedometer. In any case, it is very difficult to check a speedometer accurately at these higher speeds.

CHAPTER XIV

ELECTRICAL FAULT-TRACING

THE electrical system plays such an important part in the modern car that a comprehensive analysis of the system is an essential adjunct to the tracing of many engine, ignition, instrument, and accessory faults covered in preceding chapters. Normally, to carry out electrical fault-tracing effectively, accurate testing instruments are required, including a sensitive voltmeter, ammeter, ohmmeter, and " megger "; while bench repairs and tests of dynamo and starter armatures call for a useful device known as a "growler."

Since the average owner is unlikely to possess such equipment, and the mechanic who has access to a comprehensive range of electrical testing and repair equipment must be assumed to possess sufficient knowledge to use it correctly, this chapter is devoted to fundamental tests that require the use of only one or two instruments, but that are sufficient to enable troubles to be diagnosed with reasonable certainty. Straightforward repairs can then be carried out where possible, or the faulty component can be handed over to an electrical specialist for attention.

When carrying out any work on the electrical system it is essential to check that the polarity is correct—that is, that the correct terminal of the battery is earthed. For about thirty years, British car manufacturers adopted the positive-earth system, whereas the majority of European and American manufacturers earthed the negative terminal of the battery. From 1965 onwards, however, some British models used negative-earthing and this system was progressively adopted on new models, with the object of achieving uniformity in this respect with manufacturers in other countries, thus simplifying servicing and spares arrangements.

If any equipment which incorporates semi-conductor devices—transistors and diodes—is connected with incorrect polarity, the semi-conductors will be instantly damaged beyond repair. The installation of transistorized car radios, alternators which have diodes in the control unit, transistorized ignition equipment and so on, therefore, calls for special care.

In order to ensure that the ignition coil is correctly connected, the markings "CB" and "SW" on the terminals are replaced by "+" and "−." Another point to bear in mind is that permanent-magnet electric motors will operate on either polarity but when the polarity is reversed, the direction of rotation is also reversed and this will obviously upset the operation of windscreen wipers, interior-heater blowers and electrically-driven windscreen washers.

In conjunction with the individual tests described in the Charts, a periodical "tune-up" of the electrical system is advisable. The following series of tests, based on the practical experience of one of the largest manufacturers of electrical equipment, is well worth carrying out at 5,000 mile intervals.

Check with a hydrometer the specific gravity of each cell of the battery. Add distilled water, if necessary, to bring the electrolyte level above the plates about $\frac{3}{8}$ in. Add the water after taking the hydrometer reading.

Correct the hydrometer reading for temperature by adding four points (0·004) to the reading for every ten degrees above 60° F. Subtract four points for every ten degrees below 60° F. If the corrected reading is below 1·225, the battery should be removed for a recharge. If it is above 1·225, do not be content with the hydrometer reading alone. The hydrometer reading discloses merely the chemical condition of the battery. Additional information on the actual operating condition of the battery is needed to get a true picture of the battery's ability to deliver current under operating load.

Check each cell under operating load, when the starter motor is cranking the engine and drawing from 150 to 250 amperes from the battery. With the ignition switch off, or the high tension lead removed from the ignition coil to prevent the engine from starting, operate the starter motor and quickly check each cell of the battery with an accurate, low reading voltmeter. If the voltage of any cell falls below 1·5–1·7 volts at 80° F., or if there is a difference between cell readings of more than $\frac{1}{4}$-$\frac{1}{2}$ volt on an ordinary voltmeter, or $\frac{1}{10}$ volt on a sensitive instrument, battery trouble is indicated and the battery should be removed from the car for further check.

CHART No. 30

BATTERY TROUBLES

Symptom	Cause	Test	Remedy
Battery apparently discharged.	Undercharging; sulphated plates; shedding of active material from plates; inter-cell leakage; buckled plates; loose inter-cell connectors; individual cells faulty; self-discharge. For individual faults see below.	Test each cell with hydrometer as described in text. Test individual cells with voltmeter under high-discharge conditions (see text).	If hydrometer and voltage tests are satisfactory, check charging rate and charging circuit. If battery tests are unsatisfactory, check for individual faults listed in "Cause" column.
Sulphated plates.	Undercharging.	Check generator charging rate and voltage regulator setting (Chart No. 31).	Adjust charging rate or remedy generator and charging circuit defects.
	Battery left discharged or idle for long periods without freshening charge.		Place battery on charge at very low charging rate for long period. Renewal of plates is generally necessary if trouble is far advanced.
	Low electrolyte level, due to failure to top up battery; excessive gassing due to overcharging.	Examine electrolyte level and check charging rate.	As above.
Frequent topping-up required.	Overcharging, causing excessive gassing of electrolyte.	Check charging rate.	Adjust charging rate.
	Leakage from cells due to cracked case (see below).	Examine battery for signs of external leakage. Acid stains on metal and upholstery may be neutralized by household ammonia.	Renewal of battery is only cure; if battery is new, fitting existing plates to fresh case may be worthwhile.
Hydrometer readings on different cells vary by more than ten points on hydrometer scale.	Partial short-circuits in low-reading cells. Loss of acid by spraying or leakage from low-reading cell. Excessive water evaporation in high-reading cell.	Check cell voltages. If within 1/10 volt, re-adjust specific gravity of electrolyte in low-reading cells. If voltages differ by more than 1/10 volt, check for faults as below.	Take action indicated by results of tests. As trouble is generally due to faulty cells, renewal of battery or re-plating will probably be necessary.

Voltage readings differ on individual cells.	Defective plates in cells, due to sulphation or shedding of active material. Short-circuits in cells (see below). Open circuit in cells.	When voltmeter test indicates trouble the battery must be opened up for inspection.	Replating of defective cells may be feasible, but remainder of cells are probably also in poor condition; renewal of battery is usually the most economic course.
Difference between hydrometer and voltage readings on adjacent pairs of cells.	Inter-cell leakage, due to cracked or porous battery partition.	Remove inter-cell connector bar and test with mains voltage applied to cell posts with 60 watt lamp in circuit. If lamp lights up or glows, even faintly, leakage is taking place.	Renewal of battery case may be justified if battery is relatively new. Otherwise fit new battery.
Shedding of active material from plates.	Overcharging, resulting in excessive gassing and a low electrolyte level. Remainder of plate is exposed to concentrated acid solution and also worked doubly hard, causing the active material to break up.	Check generator charging rate, and voltage regulator setting.	Re-adjust charging rate. Battery is seldom repairable when trouble is far advanced.
Short-circuits in cells.	Buckled plates or charred separators due to overcharging. Accumulation of displaced active material short-circuiting plates.	See "Shedding of Active Material."	As battery is usually in poor condition generally, a new battery is advisable.
Reversed polarity.	If in one cell, is caused by short-circuit in cell (see above). If polarity of battery is reversed, battery has been incorrectly connected to charging circuit on bench or in car.	Verify that battery is correctly fitted to car. On most modern cars the *positive* battery terminal is earthed.	Battery is generally ruined, but may sometimes be brought back into condition by long charge at 20-hour rate on correct polarity.
Self-discharge when idle. (Slight self-discharge over long period is normal.)	If battery is on car, possible cause is leakage of current due to partial short-circuit in vehicle wiring. Discharge with battery removed from car may be caused by impurities in electrolyte, short-circuits in cells or inter-cell leakage (see above).	Disconnect live battery lead from battery. Connect sensitive voltmeter between live terminal and lead. If current flows with all car circuits switched off, fault is external to battery. If no current flows, fault is in battery.	Rectify wiring circuit, or battery troubles, as indicated by test.

CHART No. 31

LIGHTING SYSTEM TROUBLES

Symptom	Cause	Test	Remedy
Lamps give insufficient illumination.	Battery discharged or defective.	Determine whether battery is at fault by reference to Chart No. 30.	Charge battery either by a long period of daytime running or from independent electrical supply. Fit new battery.
	Lamps out of alinement or bulbs out of focus.	Inspection of lamps will reveal any defects.	Aline lamps and focus bulbs.
	Bulbs discoloured through use, or reflectors dirty.	If bulbs are blackened, check dynamo circuit for excessive voltage and battery and terminals for high resistance.	Fit new bulbs or clean reflectors.
	Reflector, lamp body or mounting improperly earthed.	Connect earthing lead between reflector and bright metal on chassis.	If lamp brilliance improves, clean all surfaces in contact, particularly between lamp and mounting and mounting and chassis.
Lamps light when switched on, but gradually fade out.	Battery discharged or defective.	Check battery as in Chart No. 30.	Charge battery either by a long period of daytime running or from independent electrical supply. Fit new battery.
Brilliance varies with speed of car.	Battery discharged.	Check battery as in Chart No. 30.	Charge battery either by a long period of daytime running or from independent electrical supply. Fit new battery.
	Excessive resistance in circuit.	Test for high resistance in circuit as described in text.	Tighten connexions, or replace faulty cables.
Lights flicker.	Loose connexion.	Systematic check of circuit is required.	Locate loose connexion and tighten.
Failure of lights.	Fuse blown.	Examine wiring for faulty cables and remedy.	Fit replacement fuse.

At temperatures below 80° F., the cell voltages will normally drop to lower values, owing to the lowered efficiency of the battery, as well as to the greater current drawn by higher cranking torque. At sub-freezing temperatures the current required will approach 300-400 amperes.

If the battery is satisfactory, check the cable leads and connexions to determine whether they are in good condition, without excessive resistance. Excessive resistance produces

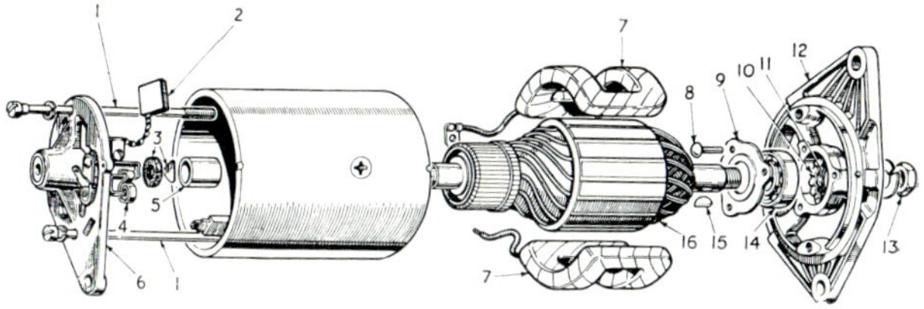

FIG. 116. TYPICAL DYNAMO TROUBLES

1. Through-bolts loose
2. Brushes worn or sticking
3. Felt ring and aluminium sealing disc require lubrication
4. Brush spring weak
5. Bearing bush worn or dry
6. Commutator end bracket damaged
7. Field coils defective
8. Rivet loose
9. Bearing retainer plate distorted
10. Corrugated washer damaged
11. Felt washer worn
12. Driving end bracket damaged
13. Pulley retainer nut loose
14. Bearing worn
15. Woodruff key loose
16. Armature windings defective. Commutator dirty, burnt or scored. Spindle bent

abnormal voltage drop, which may lower the voltage at the starter motor to such a low value that normal operation will not be obtained.

A sensitive voltmeter is required to check voltage drop. As the average owner is unlikely to possess a meter which will read to within $\frac{1}{10}$ volt, tests should be confined to a careful inspection of all leaks and connexions.

If excessive resistance is found in any of the circuits, disconnect the cables, thoroughly clean the connexions to bright metal and reconnect the cables.

Replace with new cable if the original cable appears frayed, corroded, or has broken strands. Always replace with cable at least as large as the original. Do not forget the engine earthing strap as this is part of the battery-starter motor circuit.

One more point—always check for tightness the hold-down

CHART No. 32

DYNAMO AND CHARGING CIRCUIT FAULTS

Symptom	Cause	Test	Remedy
Generator does not charge under any conditions.	Open-circuited field. Connexions unsoldered. Winding may have been broken by accident or rough usage.	Disconnect field lead on third brush and measure field current. If field circuit is open, the current will be zero. Remove tape from connexion between coils and test each coil separately.	If break is in external leads, it can be repaired. If break is on inside of field coil, install a new coil.
	Field burned out. Generator has been operated on open charging circuit or with a high resistance in the charge circuit between generator and voltage regulator, which causes the voltage to rise abnormally high, thus increasing the current through the field winding.	Examine field winding for evidence of over-heating.	Install new field coils. Make sure all connexions are tight and voltage regulator is functioning.
	Open-circuit in external field wiring. Loose connexion or break in wire from generator; field terminal to regulator field terminal; inoperative regulator.	Earth field terminal in generator frame by short wire; if generator then charges, external field circuit is faulty. Remove lead from "F" terminal on regulator and earth against bare spot on dash; if generator charges, trouble is in regulator; otherwise, trouble is in wiring or loose connexion at terminal.	Tighten connexions, replace broken wire or take unit to service station for regulator repair.
Generator will not charge until relay contacts are closed by hand.	Dirty or greasy commutator, due to improper brushes; excessive lubrication of commutator end bearing; cover band left off generator.	Examine commutator; commutator may be covered with a thin film or glaze which is not easily distinguished.	Clean commutator with a rag moistened with petrol. To polish commutator, use a strip of fine (No. 00) sandpaper (do not use emery.) Pass this around commutator with sanded side in and draw it back and forth while the armature is rotating.
Charging rate low or unsteady; arcing and burning of brushes.	High mica. Copper is softer than mica and therefore it wears away more rapidly, leaving the mica projecting.	Examine the commutator closely. The mica should be below the surface of the bars.	True up the commutator and undercut the mica.

Generator stops charging, or charges unsteadily, at high speed.	Commutator out of round, caused by one or more bars higher or lower than the rest. Commutator may have been turned down by a careless workman.	Rotate the armature slowly, and hold a pencil on one of the brushes. If the commutator is out of round, the brush can be felt to rise and fall as the armature rotates.	Have commutator turned down in a lathe. Rotate the armature at high speed and take a very light cut with a sharp tool. Do not remove any more material than is necessary to true up the commutator. Polish the commutator with a strip of fine (No. 00) sandpaper. Mica between commutator bars should now be undercut to a depth of 1/32 in.
Charging rate low or unsteady; arcing at brushes.	Brush spring weak, caused by loose or missing brush lead, allowing current to flow through brush spring. Excessive arcing of brushes causing overheating of springs. Improper assembly of spring.	Test spring—tension should be 22 to 26 ounces for main brush and 16 to 20 for third brush, but obtain figures from manufacturer if possible.	New spring. Make sure that brush leads are in place.
	Brush sticks in holder due to bent or dirty holder.	Examine brush and holder to make sure brush slides freely.	Clean brush and holder. Straighten holders.
	Brush ring earthed, due to broken or burned insulation; metal chips, dirt and carbon dust on brush ring.	Raise brushes from commutator and test between insulated brush and generator head with a test lamp.	Thoroughly clean the brush holder and brush ring insulation. If the insulation is damaged, it may be necessary to replace the insulating bushings in the generator head.
Low or zero charging rate; commutator burned at the bars connected to open-circuited coils.	Open-circuited armature; connexions unsoldered at commutator; wire broken.	Test armature as for "Shorted Armature," below.	Re-solder connexion. If damage is not accessible, exchange armature for new one.
	Earthed armature wires come in contact with armature shaft or laminations, due to damaged or defective insulation.	Insulate brushes from commutator and test between commutator and shaft with test lamp. To locate the exact position of the earth it may be necessary to disconnect the leads from the armature and test each coil separately. This should only be done by an experienced armature repair man.	If the defect is readily accessible, it can be re-insulated, otherwise the armature should be exchanged for a new one.

CHART 32 (continued)

Symptom	Cause	Test	Remedy
Armature overheats; solder melted out of commutator; insulation burned; charging rate low or zero; may be intermittent or "swinging short."	Shorted armature. Wires come into contact with one another, due to damaged or defective insulation. May also be due to metal chips between commutator bars.	With generator removed from engine, connect generator to battery with ammeter in circuit. Generator should operate as a motor and ammeter will indicate the current drawn. If the armature jerks and draws a pulsating current, it is shorted. Make sure that the armature is free to rotate and that it does not rub on the poles or bind in the bearings.	Clean out slots in commutator bars. If the short circuit between wires is not readily accessible, the armature should be exchanged for a new one.
Generator noisy; armature cannot be easily rotated.	Armature rubs or binds, due to loose pole; worn bearings; bent shaft; improper assembly.	Examine generator; look for scored places on armature core.	Tighten loose parts. Make sure heads are properly fitted against end of frame. Replace worn bearings.
High generator voltage irrespective of voltage regulator setting. Burned-out lamp bulbs.	Earthed field, due to damaged or defective insulation; metal particles between field coil and frame or pole.	Disconnect field lead at third brush and generator field terminal and test with lamp between field terminal and starter terminal. If the lamp lights, the field is earthed. To locate damaged coil, disconnect coils and test each one separately.	Re-insulate or replace field winding.
Noisy generator. Bearings heat or burn out.	Shaft bent or sprung. Loose pole, rough usage or thrown windings.	Rotate armature on its shaft centers. A bent shaft can be detected with an indicator which should not show more than 0·002 in. eccentricity.	Sometimes a bent shaft can be straightened but usually a new armature will be required.

clamps, but avoid drawing them down so tightly as to distort the battery case. If the carrier is corroded, remove the

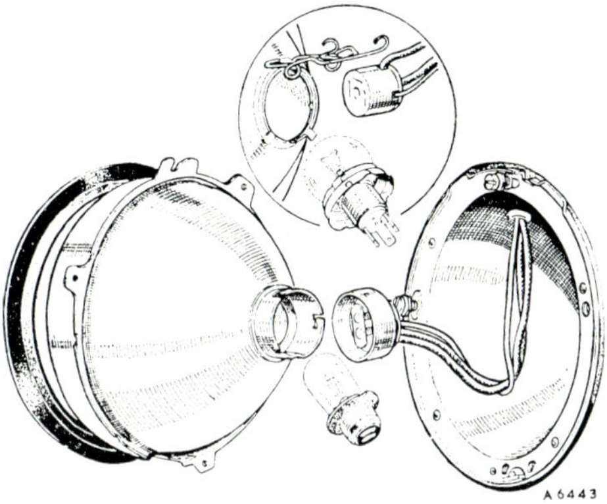

FIG. 117. HEADLAMP ASSEMBLY USING PRE-FOCUS BULB
Inset, European pattern

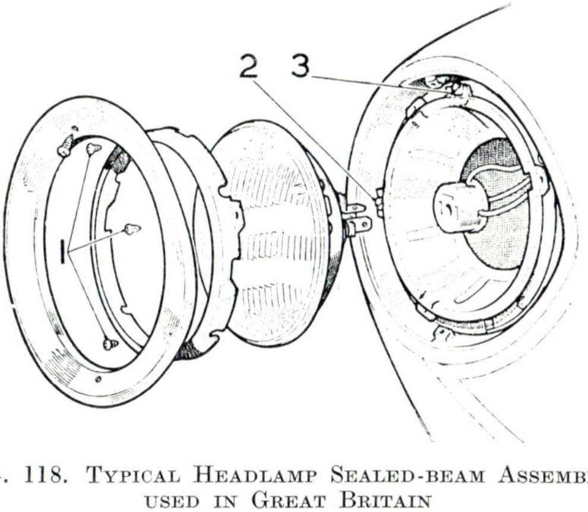

FIG. 118. TYPICAL HEADLAMP SEALED-BEAM ASSEMBLY USED IN GREAT BRITAIN

1. Screws retaining reflector assembly. 2. Horizontal beam adjusting screw. 3. Vertical beam adjusting screw

battery and clean the carrier with a wire brush, then paint with acid-proof paint.

After the battery and cables have been checked, and all

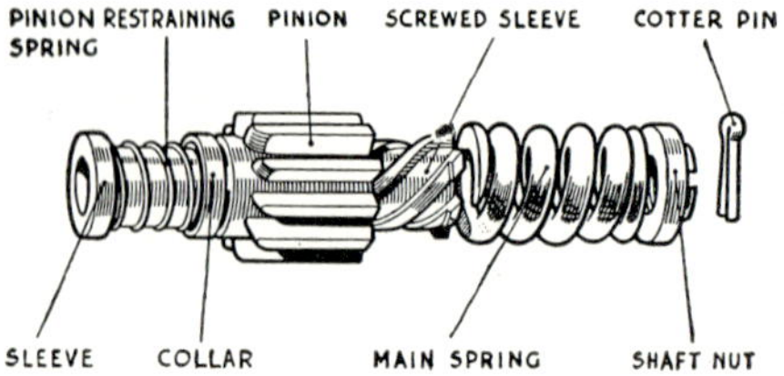

FIG. 119. COMPONENT PARTS OF A TYPICAL BENDIX DRIVE

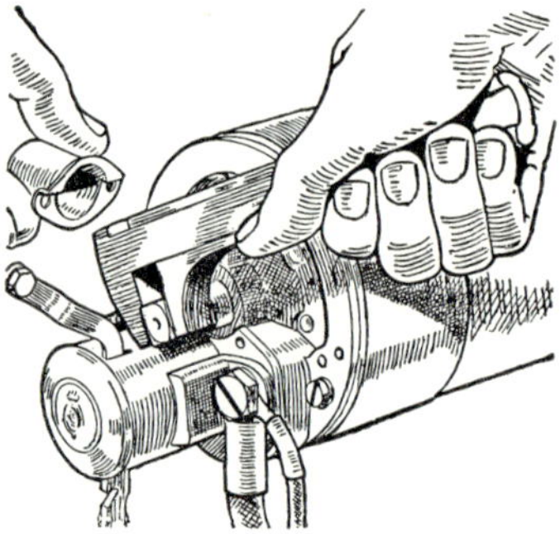

FIG. 120. FREEING A JAMMED STARTER PINION

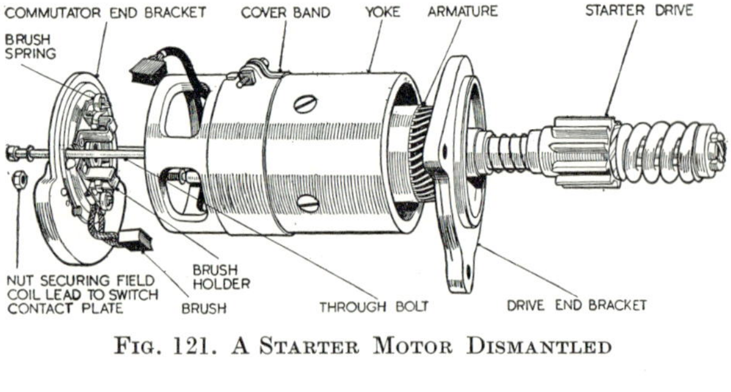

FIG. 121. A STARTER MOTOR DISMANTLED

necessary corrections made, the next checks should be made on the starter motor system, which consists of the battery and cables, already checked, and the starter motor. In checking the starter motor, examine the brushes for wear

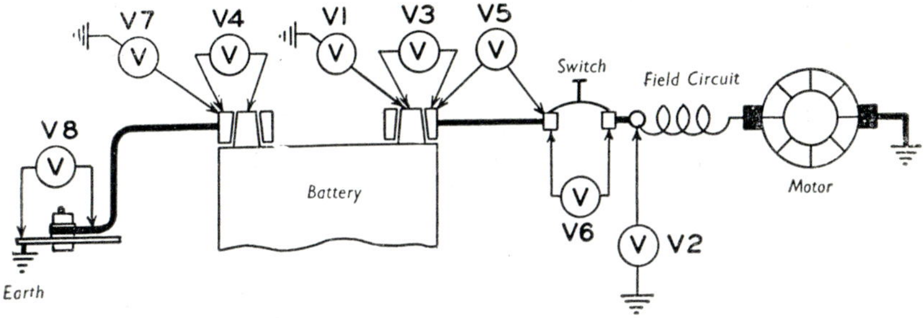

FIG. 122. THE STARTER CIRCUIT SHOULD BE TESTED FOR CONTINUITY OR HIGH RESISTANCE WITH A SENSITIVE VOLT-METER, IF AVAILABLE, OR A TEST LAMP IN THE ORDER SHOWN, FROM V1 TO V8

and for freeness of the holders on their pivots. The brush spring tension should be sufficient to give good firm contact of the brushes on the commutator. Check the pigtail connexions and the brush screws for tightness. Replace worn brushes.

Inspect the commutator for roughness, burned spots, and high mica. If necessary, dismantle the armature from the unit, put it in a lathe and take a light cut off the commutator. If high mica is found, the indication is that the mica is hard and it should be undercut.

The Dynamo and Charging System. Attention can now be turned to the generator and regulator system. First, loosen the generator mounting so that the pulley is free, and note the sideplay and the freeness of rotation, to determine the condition of the bearings. If the bearings are worn, or excessively tight, remove the generator and replace or clean the bearings. Add 8 to 10 drops of light engine oil to the lubricator. Avoid over-oiling as this will cause gumming of the commutator and reduced output.

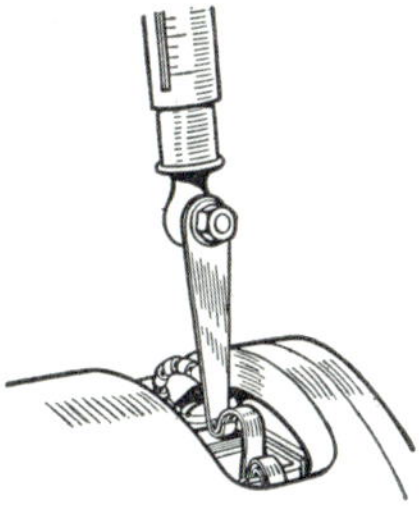

FIG. 123. USING A SPRING BALANCE TO CHECK THE TENSION OF THE DYNAMO BRUSH SPRINGS

CHART No. 33

STARTER MOTOR TROUBLES

In considering starter motor troubles, it is assumed that the battery is in good condition and fully charged (see Chart No. 30)

Symptom	Cause	Test	Remedy
Starter does not engage or does not release. If it does not release, the armature will be driven at high speed and the windings thrown out of the slots.	Bendix drive or over-running clutch defective or broken. Backfire; defective or damaged part in Bendix drive; weak pedal return spring; or sluggish pedal shift mechanism on pedal engagement starters.	Check operation of mechanism.	Install new drive assembly. Replace pedal return spring if weak. Free pedal shift mechanism. Check ignition timing and be sure that it is properly set and that automatic spark advance mechanism is not stuck.
	Loose or broken connexion.	Check over circuit systematically.	Locate and tighten loose connexion, or re-make broken connexion.
Motor will not crank engine or cranks it slowly.	Armature short-circuited. Damaged or defective insulation on armature.	Test armature with growler.	If damage is readily accessible, it can be repaired; otherwise exchange armature for a new one.
Motor cranks engine slowly or not at all. Poor commutation; commutator burnt.	Armature open-circuited. Connexions unsoldered at commutator.	Examine armature for evidences of loose connexions. Test with millivoltmeter if damage is not visible.	Re-solder connexions or exchange armature for new one.
Poor commutation; commutator burnt, motor cranks slowly.	Brush spring weak. Spring damaged or not properly installed.	Check brush spring tension which should be 24 to 28 ounces.	New brush spring; tighten loose connexions.

Motor runs at high speed without load but runs slowly when cranking engine.	Field winding short-circuited. Damaged or defective insulation on field coils.	Short-circuited field coils are difficult to detect due to the fact that the normal resistance of the field coil is very low. The simplest process of locating damaged field coils is by the process of elimination. Make sure that all other parts of the motor are in order then install new field coils. If motor then operates satisfactorily, the original coils were short-circuited.	
	Earthed field coil. Damaged or defective insulation.	Disconnect field leads from brushes; test between terminal and frame with 110 200-250 volt test lamp. If lamp lights, the field winding is earthed.	Re-insulate winding or install new coil.
Motor will not run. No current drawn from battery.	Field circuit open. Broken or unsoldered connexion.	Use test lamp and test between motor terminal and end of field winding which is connected to the insulated brush. If the lamp does not light, the field is open.	Re-solder connexions or install new field.
Motor turns slowly or not at all.	Tight armature shaft bearings. Lack of lubrication; bearings not properly alined; shaft damaged.	Examine motor; rotate armature by hand to determine if it moves freely.	If bearing is not damaged it can be cleaned and lubricated; otherwise new parts may be necessary. Make sure that lubricating device in commutator end is in good condition and that armature turns freely in its bearings.
Motor noisy; armature laminations damaged; armature may be shorted.	Armature rubs on pole. Bent shaft; loose pole; worn bearings.	Examine motor; remove armature and rotate it about its centers, taking indicator reading of the shaft. Eccentricity should not exceed 0·002 in.	Shaft can usually be straightened, otherwise exchange for new armature. Tighten poles.
Excessive side play in shaft; motor noisy. If bearings are worn until armature strikes pole, armature laminations may be damaged.	Bearings worn. Long use; lack of lubrication; dirt or grit in bearings.	Examine motor; side play in bearings should be very small.	Install new beads or bushings. Make sure that lubricating device in commutator end is in good condition.

Next, remove the cover band or the end-plate and inspect for thrown solder. Thrown solder indicates overheating, resulting from excessive output. Remove the generator so that the armature can be resoldered and repaired, and after re-installation on the car, reduce the generator output, to that specified by the manufacturer, by lowering the current regulator setting to the correct value.

Check the brushes for wear and freeness in their holders. The brush spring tension should be within the limits specified by the manufacturer, so that good, firm contact is established with the commutator. Check the pigtail lead connexions and the brush screws for tightness. Replace worn brushes to prevent generator failure before the next tune-up period. Be sure to use the genuine brushes specified for service. Even though some replacement brushes are mechanically interchangeable and have an appearance similar to the original equipment brushes, it does not mean they will give as satisfactory operation.

Inspect the commutator for wear, roughness, and high mica. A dirty commutator can be cleaned by holding fine sandpaper or a bedding stone against it with the generator in operation. The bedding stone, also used to seat new brushes, is an abrasive which, held against the revolving commutator, disintegrates and is carried under the brushes and seats them in a few seconds. The bedding stone is sufficiently hard to remove dirt and grease from the commutator, but will not scratch or injure it.

Blow out all dust. Never use emery cloth to clean the commutator, as emery will embed in the commutator and cause rapid brush wear. If the commutator is worn, rough, out of round, or has high mica, remove the armature and take a cut off the commutator in a lathe. Undercut the mica.

Check the belt and replace it if worn. Tighten to the correct tension, but avoid excessive belt tension, as this may cause rapid bearing wear.

Avoiding Excessive Voltage Drop. Extra resistance resulting from defective leads and poor connexions will prevent a proper charge from reaching the battery. Excessive voltage drop is an indication of extra resistance. The voltage drop between the generator and the battery may be easily measured.

Earth the "F" terminal of the generator, adjust the

generator speed so that twenty amperes are flowing to the battery, and check the generator and battery voltage. The difference between the two readings is the voltage drop.

If the voltage drop is above $\frac{3}{4}$ volt with twenty amperes flowing, excessive resistance exists in the circuit. Check

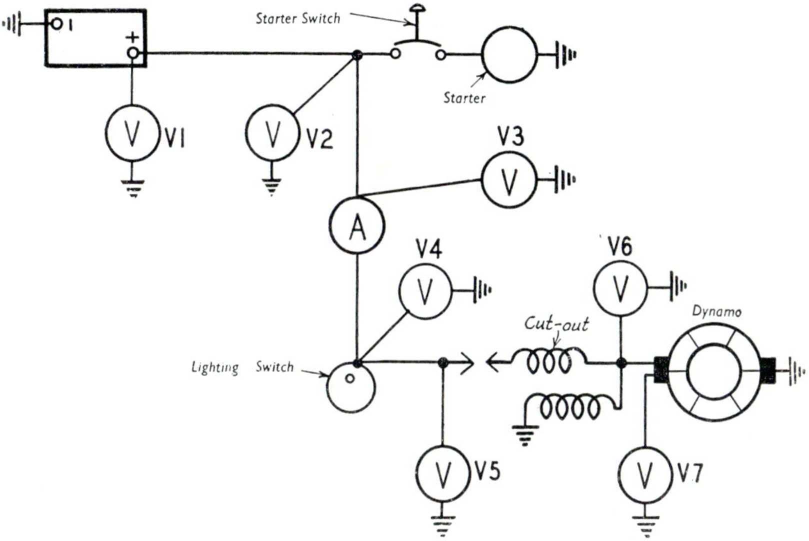

FIG. 124. SUCCESSIVE VOLTMETER TESTS, STARTING AT V1 AND ENDING AT V7, TO CHECK THE CHARGING CIRCUIT FOR CONTINUITY AND HIGH RESISTANCE. A LAMP OF BATTERY VOLTAGE CAN BE SUBSTITUTED FOR THE VOLTMETER. A NEGATIVE-EARTH SYSTEM IS SHOWN

the leads and connexions separately to locate the source of trouble. Clean and tighten the connexions and replace the leads if necessary to eliminate resistance.

On the current and the voltage regulator units, one of the most important service operations is the cleaning of the contact points. The regulator contact points, like the contact points of the ignition distributor, will not operate indefinitely without attention. It will be necessary to clean them occasionally, since dirty or oxidized contact points will cause reduced generator output and run-down batteries. To clean the points, use a spoon or riffler file. Do not use a flat file on the flat point, since it will not clean out the slight cavity formed in normal operation. Do not use emery cloth or sandpaper, since particles of sand or emery may embed in the contacts and prevent normal operation.

The setting of the regulator should not be disturbed. Adjustment must be carried out in accordance with the figures quoted in the manufacturer's schedule. Regulator checks are therefore best left to a service station specializing in the particular make of equipment.

Alternators. When an alternator is used in place of a dynamo, it is essential to obtain the appropriate servicing instructions before attempting to carry out fault-tracing or to adjust the output. The correct polarization must be carefully checked. Any attempt to re-polarize an alternator may damage the alternator itself, the regulator, when this contains a diode, or the associated circuit. Battery polarity must therefore be checked before any connexions are made or before connecting-up a battery charger. When charging the battery from an external source of current, in fact, it is advisable to disconnect it from the alternator circuit.

It should also be remembered that because an alternator charging circuit does not employ a cut-out, the battery voltage is applied at all times to the main cable connecting the alternator to the battery and to the output terminal on the alternator. Care must be taken, therefore, not to earth the cable or the terminal. The cable connected to the field terminal on the alternator should not be earthed since, if the ignition is switched on, this may damage the regulator or the relay and the wiring. Similarly, earthing the cable connected to the AL terminal on the alternator and on the warning-light control will damage the alternator.

The alternator should never be run with the battery disconnected and its field windings energized, as this will damage the control unit. Never check the rectifier diodes or transistor with an ohmmeter which incorporates a hand-driven generator.

Finally, remember that an alternator has a high output and imposes a heavy load on the fan belt, which *must* be kept correctly tensioned and must be renewed at the first sign of deterioration.

INDEX